The Malliavin Calculus

Denis R. Bell
University of North Florida

Dover Publications, Inc.
Mineola, New York

Bibliographical Note

This Dover edition, first published in 2006, is an unabridged republication of the work originally published by Longman Scientific & Technical, Essex, England, and John Wiley & Sons, Inc., New York, in 1987. A new Preface and Appendix have been especially prepared by the author for this edition.

International Standard Book Number: 0-486-44994-7

Manufactured in the United States of America
Dover Publications, Inc., 31 East 2nd Street, Mineola, N.Y. 11501

To Cindy and Justin, with love

Contents

Preface to the Dover Edition

N. Wiener provided the first rigorous mathematical description of Brownian motion in 1923, creating what is now known as the Wiener process. The importance of Wiener's work was greatly enhanced by the invention by K. Itô of a theory of stochastic integration and an accompanying theory of stochastic differential equations, in which the Wiener process plays a central role. These equations have enormous practical significance as they provide a way to model physical systems subject to the influence of noise, in the same way that ordinary differential equations are used to model deterministic systems.

In addition to their physical applications, stochastic differential equations have important mathematical ramifications. In particular, it is possible to use stochastic differential equations to obtain probabilistic solutions to the heat equation and to a large class of related partial differential equations. In developing this link between stochastic and classical analysis, one is led to study the solution of a stochastic differential equations as a function of the driving Wiener process. Typically, this function is non-smooth in the sense of classical calculus.

The Malliavin calculus, a methodology for studying the regularity of such Wiener functionals, was invented in order to address this problem. The subject derives from two fundamental papers ([19] and [20]) of P. Malliavin that appeared in the late seventies. The original purpose of these papers was to exploit the link between partial and stochastic differential equations to give a probabilistic proof of Hörmander's hypoellipticity theorem. However, it was soon realized that Malliavin's approach would have applications far beyond the scope of this problem.

When I wrote this book in 1984, the Malliavin calculus was being developed as a tool to study a wide variety of problems. In addition to the anticipated proof of Hörmander's theorem, it has since been used to obtain a host of results in areas as diverse as filtering theory, differential geometry, and mathematical finance. The theory has now matured to the point where it is part of the canon of stochastic analysis.

The content of the second edition is identical to the original, except for the addition of an appendix, in which a topic treated in an earlier section (i.e. the relationship between quasi-invariance and admissibility of measures) is developed further. It is stated in the original preface "This material is obviously not intended to be an exhaustive account of the field to date". Clearly, this is even more true now. The reader is encouraged to consult the works listed below to learn about some of the developments that have occurred in the two decades since the original edition of the book was published.

Denis Bell
Jacksonville, 2005

Supplementary Literature

Bell, D, *Degenerate stochastic differential equations and hypoellipticity*, Pitman Monographs and Surveys in Pure and Applied Mathematics, 79. Longman, Harlow, 1995.

Bismut, J. M., *Large Deviations and the Malliavin calculus*, Progress in Mathematics, vol. 45. Birkhauser, Boston, 1984.

Malliavin, P. *Stochastic analysis.* Grundlehren der Mathematischen Wissenschaften, 313. Springer-Verlag, Berlin, 1997.

Nualart, D., *The Malliavin calculus and related topics.* Probability and its Applications. Springer-Verlag, New York, 1995.

Watanabe, S., *Lectures on stochastic differential equations and Malliavin calculus.* Tata Institute of Fundamental Research Lectures on Mathematics and Physics, 73. Springer-Verlag, Berlin, 1984.

Conference on Applications of Malliavin Calculus in Finance, *Math. Finance* 13, (2003), no. 1. Blackwell Publishers, Inc., Malden, MA, 2003.

In retrospect, the statement on the cover of the first edition "Enough technical background is included to make the material accessible to readers without specialized knowledge in stochastic analysis" was perhaps a trifle optimistic! An excellent source for background information is Gikhman and Skorohod's book *Stochastic Differential Equations*[11].

Preface

The appearance of a new idea in mathematics is often the result of an attempt to solve a specific problem. However, sometimes this idea proves to be so novel and interesting that it transcends the problem that it was originally designed to solve and becomes an area of study in its own right. Such is the case with the stochastic calculus of variations developed by Paul Malliavin in 1976 in order to provide a probabilistic proof of Hörmander's theorem, which has since come to be known as the Malliavin calculus.

In its original form, the Malliavin calculus is a beautiful but highly elaborate complex of ideas combining deep results from probability theory and functional analysis. Since its inception it has undergone considerable simplification and extension by a number of mathematicians and is now a powerful tool for proving a variety of results. The theory admits many different treatments, and two of the main contributors to the field, Stroock and Bismut, have developed the subject in different and non-equivalent directions.

This book is intended to serve as an introduction to the Malliavin calculus, suitable for non-specialists who wish to learn about the subject.

Chapter 1 contains the technical background needed in later sections. In chapters 2 and 3 detailed accounts of the work of Stroock and Bismut are given. For the sake of clarity, these versions of the subject are presented in their original context, i.e. as a tool for proving Hörmander's theorem. In chapter 4 we obtain Malliavin's result via an alternative, elementary method. In order to complete the proof of Hörmander's theorem, it is necessary to study the invertibility of a certain stochastic matrix which appears in the earlier chapters. This is done in chapter 6, where the results of Norris are presented. Finally, in chapter 7 we briefly describe some of the applications of the Malliavin calculus to problems other than Hörmander's.

This material is obviously not intended to be an exhaustive account of the field to date. However, we hope that it will serve its purpose as an introductory work, while at the same time giving some unity to a subject that even now is expanding in many different directions.

I would like to express my gratitude to David Elworthy for introducing me to the Malliavin calculus, for providing guidance and encouragement during my time as a graduate student at the University of Warwick, and for his critical review of the manuscript which resulted in many improvements. I am also indebted to Dan Stroock for sharing his expertise with me on several occasions. Finally, I would like to thank Bridget Buckley at Pitman Publishing and the staff at Longman for their courtesy and for their efficient handling of the publication of this book.

Denis Bell
Boston, 1986

Introduction

We will start by describing some of the ideas that motivated the development of the Malliavin calculus.

Suppose that $A_0, A_1, \ldots, A_n$ are smooth vector fields and f is a real-valued bounded continuous function, defined on $\mathbf{R}^d$. Let $\mathcal{G}$ denote the second order differential operator

$$\mathcal{G} \equiv \frac{1}{2}\sum_{i=1}^{n} A_i^2 + A_0. \tag{0.1}$$

An important problem in the theory of partial differential equations is the determination of conditions on $A_0, \ldots, A_n$ under which the Cauchy problem

$$\left.\begin{aligned} &\frac{\partial U}{\partial t} = \mathcal{G}U \\ &U(t, \cdot) \to f \quad \text{as} \quad t \downarrow 0 \end{aligned}\right\} \tag{0.2}$$

admits a smooth fundamental solution. By this is meant a real-valued function P defined on $(0, \infty) \times \mathbf{R}^{2d}$ with $P(t, \cdot)$ smooth on $\mathbf{R}^{2d}$ for every $t \in (0, \infty)$, such that the map

$$U(t, x) \equiv \int_{\mathbf{R}^d} P(t, x, y) f(y)\, \mathrm{d}y, \qquad (t, x) \in (0, \infty) \times \mathbf{R}^d$$

satisfies (0.2).

Let

$$A_i = \sum_{j=1}^{d} a_{ji} \frac{\partial}{\partial x_j}$$

for each $i = 0, \ldots, n$ and define A to be the matrix $(a_{ji})_{j,i=1}^{d,n}$ and A_0 to be the vector $(a_{j0})_{j=1}^{d}$. It had been known for some time that a smooth fundamental solution to (0.2) exists in the *elliptic* case, where AA^* is everywhere invertible. However, in a ground-breaking paper[14] of 1967, Lars Hörmander proved that this holds under a considerably weaker hypothesis. Let

$$[V, W](x) \equiv DV(x)W(x) - DW(x)V(x)$$

denote the Lie bracket of two C^1 vector fields V and W. Hörmander showed that a smooth fundamental solution exists under the assumption that the vectors

$$\{A_i, [A_j, A_k], [[A_j, A_k], A_l], \ldots; 1 \le i \le n,\quad 0 \le j, k, l, \ldots \le n\} \tag{0.3}$$

span $\mathbf{R}^d$ at each point.

There is an intimate connection between the above Cauchy problem and the theory of stochastic differential equations. Let w denote normalized n-dimensional Brownian motion and consider the family of (Stratonovich) stochastic differential equations

$$\xi_t^x = x + \int_0^t A(\xi_s^x)\,\mathrm{d} \circ w_s + \int_0^t A_0(\xi_s^x)\,\mathrm{d}s; \qquad (t, x) \in [0, \infty) \times \mathbf{R}^d \tag{0.4}$$

Then for each $x \in \mathbf{R}^d$, $\{\xi_t^x\}_{t \ge 0}$ is a Markov process. Let $\{P_t\}_{t \ge 0}$ be the semi-group defined on bounded continuous functions g on $\mathbf{R}^d$ by

$$(P_t g)(x) = E[g(\xi_t^x)], \qquad t \ge 0 \tag{0.5}$$

where E denotes expectation, and define $U(t, x) \equiv (P_t f)(x)$. If $U(t, x)$ is C^2 in x for each $t > 0$, then it satisfies (0.2). To see this argue as follows: an application of the Itô lemma shows that $\mathcal{G}$ is the infinitesimal generator of $\{P_t\}$ (defined on bounded C^2 functions). The semi-group property then gives

$$\frac{\partial U}{\partial t} = \frac{\partial}{\partial t}(P_t f) = \mathcal{G}(P_t f) = \mathcal{G} U.$$

The second part of (0.2) follows from the a.s. continuity of ξ^x. In particular since

$$(P_t f)(x) = \int_{\mathbf{R}^d} P(t, x, \mathrm{d}y) f(y)$$

where $\{P(t, x, \mathrm{d}y)\}$ denotes the family of transition probabilities for $\{\xi_t^x\}$, it can be shown that the existence of a smooth fundamental solution to the Cauchy problem is equivalent to the statement that for each positive t and $x \in \mathbf{R}^d$ the measure $P(t, x, \mathrm{d}y)$ is absolutely continuous with respect to the Lebesgue measure on $\mathbf{R}^d$, and the corresponding family of densities $\{P(t, x, y)\}$ can be chosen to be smooth in (x, y).

Thus Hörmander's theorem can be formulated as the statement that under condition (0.3) there exists a family of smooth transition densities for the solution of equation (0.4).

Before Malliavin's papers,[19,20] no direct method existed for proving this; indeed the problem had proved to be intractable to standard probabilistic techniques. The reason for this can be understood by considering the following example which illustrates the usual procedure for obtaining such regularity results. Let X be a random variable with a smooth density ρ and F a smooth real-valued function on $\mathbf{R}$ with a non-vanishing derivative. Suppose that one wishes to study the measure induced by the random variable $Y = F(X)$. Then for every test function φ on $\mathbf{R}$ and positive integer k, successive integration by parts yields

$$E[\varphi^{(k)}(Y)] = \int_{\mathbf{R}} \varphi(t)(R^k\rho)(t)\,\mathrm{d}t \tag{0.6}$$

where R^k is the kth iterate of the operator R, defined by $R\rho = -(\rho \circ F^{-1}/F' \circ F^{-1})'$. If each of the functions $R^k\rho$ is integrable with respect to the Lebesgue measure, then estimating the integral in (0.6) in an obvious way gives

$$|E[\varphi^{(k)}(Y)]| \le \|\varphi\|_\infty \, \|R^k\rho\|_{L^1}, \qquad k \in \mathbf{N} \tag{0.7}$$

A result from harmonic analysis (Lemma 1.14) now implies that Y is absolutely continuous and has a smooth density.

It is possible to extend this argument to the case where X and Y take values in higher dimensional vector spaces (cf. Theorem 5.1). Unfortunately the following essential difficulty is encountered if one tries to apply it to the study of the measure induced on $\mathbf{R}^d$ by the random variable ξ_t^x. The function here which corresponds to F in the above example is the map defined on the paths w of Brownian motion by $w \to \xi_t^x$, and this proves to be highly irregular in the sense of classical differential calculus. In particular it lacks the smoothness required to implement an integration by parts scheme of the type described above. The problem requires a calculus which is compatible with the transformations defined by stochastic integration, and this is exactly what Malliavin provided.

The key ingredient in Malliavin's work is the introduction of a symmetric linear operator $\mathscr{L}$ defined on a dense subspace of the Hilbert space of L^2-Wiener functionals, together with a bilinear form $\langle\,,\rangle$ defined by

$$\langle f, g\rangle \equiv \mathscr{L}(fg) - f\mathscr{L}g - g\mathscr{L}f; \qquad f, g \text{ in the domain of } \mathscr{L}.$$

These operations may be applied arbitrarily often to the solution map of equation (0.4)† and satisfy the condition that if f and g are in the domain

† In order to simplify the notation, we assume here that we are dealing with a real-valued stochastic differential equation.

of $\mathscr{L}$, and φ is a C^1 function, then $\varphi \circ f$ is also in the domain of $\mathscr{L}$ and

$$\langle \varphi \circ f, g \rangle = \varphi' \circ f \langle f, g \rangle. \tag{0.8}$$

The operators $\mathscr{L}$ and $\langle , \rangle$ are manipulated as follows to generate the estimates in (0.7). Suppose that φ is a test function as in (0.7). It can be shown that Hörmander's condition ((0.3)) implies that the random variable $\langle \xi_t^x, \xi_t^x \rangle$ is almost everywhere non-zero, and the symmetry of $\mathscr{L}$, together with (0.8), allow one to write

$$\begin{aligned} E[\varphi'(\xi_t^x)] &= E\left[\frac{\langle \varphi \circ \xi_t^x, \xi_t^x \rangle}{\langle \xi_t^x, \xi_t^x \rangle}\right] \\ &= E\left[\frac{\mathscr{L}(\varphi \circ \xi_t^x \cdot \xi_t^x) - \varphi \circ \xi_t^x \mathscr{L}(\xi_t^x) - \xi_t^x \mathscr{L}(\varphi \circ \xi_t^x)}{\langle \xi_t^x, \xi_t^x \rangle}\right] \\ &= E\left[\varphi \circ \xi_t^x \left\{ \xi_t^x \mathscr{L}\left(\frac{1}{\langle \xi_t^x, \xi_t^x \rangle}\right) - \frac{\mathscr{L}(\xi_t^x)}{\langle \xi_t^x, \xi_t^x \rangle} - \mathscr{L}\left(\frac{\xi_t^x}{\langle \xi_t^x, \xi_t^x \rangle}\right)\right\}\right]. \end{aligned}$$

The first order (i.e. $k = 1$) estimate in (0.7) can be obtained from this by showing that the random variable contained in the bracket $\{ \}$ is integrable. Iteration of this procedure leads to the higher order estimates. Existence and smoothness of a density for ξ_t^x then follow as before. The details of this argument can be found in chapter 2.

The layout of the book is as follows. Chapter 1 contains the technical background that will enable a reader with no specialized knowledge of stochastic analysis to follow the material presented in chapters 2–7. In section 1.1 we describe the construction of Gaussian measures on infinite dimensional vector spaces and introduce the Wiener measure (the law of Brownian motion) which plays a central role throughout. Section 1.2 contains a brief introduction to the theory of stochastic integration with respect to Brownian motion. We would like to point out that although adequate for the present purpose this discussion is highly rudimentary in nature. A reader who wishes to learn the subject of stochastic integration is advised to consult Gikhman and Skorohod,[11] where he will find an excellent account of the theory of stochastic integration with respect to Brownian motion, and McShane[21] for a more general treatment. In section 1.3 we state some results that are required in later chapters, giving references where proofs may be found.

Chapter 2 is a description of the functional analytic development of the Malliavin calculus due to Stroock. In section 2.1 an axiomatized version of Malliavin's operator $\mathscr{L}$ is defined. Section 2.2 describes a construction of this operator based upon Wiener's decomposition theorem (Theorem 1.16). In Section 2.3 it is shown that the operator $\mathscr{L}$ implies regularity results for measures induced by maps in the domain of $\mathscr{L}$, and in section

2.4 the machinery developed in the previous sections is used to study the transition probabilities $P(t, x, dy)$ discussed above.

In chapter 3 we present an alternative approach, devised by Bismut. This method is outlined in section 3.1. Section 3.2 contains results which assert that the solution of a stochastic differential equation depends smoothly upon its initial point. In section 3.3 the stability theorems of section 3.2 together with the Girsanov theorem (Theorem 1.15) are used to rederive the main result of section 2.4 by means of a perturbation argument.

Chapter 4 contains an elementary derivation of Malliavin's result due to the author, which was motivated by the following considerations. The structure of any Gaussian† measure on a Banach space E is determined by a certain Hilbert subspace $H \subset E$ (cf. section 1.1). In the work of Ramer,[25] who studied the transformation of Gaussian measure under a differentiable map, a central role is played by the restriction of the map to the subspace H. This observation led us to study the action of the *Itô map* $g: w \to \xi^x$ on $L_0^{2,1}$, the canonical Hilbert subspace for the Wiener measure.‡ It can be shown that g is the stochastic extension of a deterministic map $\tilde{g}: y \to z$ defined on $L_0^{2,1}$ by the integral equation

$$z_t = x + \int_0^t A(z_s) y_s' \, ds + \int_0^t A_0(z_s) \, ds, \qquad t \in [0, 1].$$

The situation is illustrated by the following diagram

$$\begin{array}{ccc} C_0 & \xrightarrow{g} & C_x \\ \uparrow{\scriptstyle i} & & \uparrow{\scriptstyle i} \\ L_0^{2,1} & \xrightarrow[\tilde{g}]{} & L_x^{2,1} \end{array}$$

It turns out that $\tilde{g}$ has the analytic structure which g lacks, and this makes it possible to derive the estimates in (0.7) by integrating by parts with respect to the maps $\tilde{g} \circ P_m$ and then applying a limiting argument, where $\{P_m\}_{m \geq 1}$ is a sequence of projections from C_0 into $L_0^{2,1}$ converging strongly to the identity on $L_0^{2,1}$.

In chapter 4 we implement a modified form of this scheme using a sequence of piecewise linear approximations to g which imitate the construction of the Itô integral (cf. section 1.2). These approximations (defined in section 4.2) finite dimensionalize the differential analysis in

† The Wiener measure is an example of a Gaussian measure on an infinite dimensional Banach space.

‡ Here w and ξ^x are as in equation (0.4). See section 1.1 for the definitions of the spaces $L_0^{2,1}$ and C_0.

the problem. The main results of this chapter are contained in section 4.4. Section 4.3 deals with the derivation of the covariance matrix σ which appears in the statements of the theorems in section 4.4 (see below).

Section 5.1 contains a brief account of Malliavin's paper.[19] In section 5.2 we discuss the relationship between the material contained in the previous chapters and give a condition under which the results of Stroock and Bismut are shown to be equivalent. In section 5.3 a relationship is developed between a certain differential operator (which is closely related to Malliavin's operator $\mathscr{L}$) and transformation results such as Ramer's theorem.

The proof of the probabilistic form of Hörmander's theorem (Theorem 6.1) is completed in chapter 6. The earlier application of the Malliavin calculus to equation (0.4) generates a matrix-valued process $(\sigma_t)_{t\geq 0}$ and the existence of smooth† densities for the transition probabilities $P(t, x, dy)$ is shown to depend upon non-degeneracy properties of the matrices σ_t. It turns out to be sufficient to show that, under Hörmander's condition, σ_t is invertible for each positive t and $\sigma_t^{-1} \in \bigcap_{p\geq 1} L^p$. In section 6.1 we give a geometric interpretation of Hörmander's condition in terms of the original stochastic differential equation and show that it implies the invertibility of σ_t. The required L^p estimates on σ_t^{-1} are obtained in section 6.2.

In chapter 7 we describe some applications of the Malliavin calculus, for the most omitting proofs of subsidiary lemmas. Section 7.1 deals with an application to the non-linear filtering problem, due to Michel. In section 7.2 we discuss some work of Stroock in which he uses the Malliavin calculus to study a problem that occurs in statistical mechanics concerning an infinite particle system with local interactions. Finally, in section 7.3 we prove that the measures induced by a class of stochastic differential equations with values in a Hilbert space possess a certain differentiability property, thereby generalizing the result proved in the earlier chapters to infinite dimensions.

† We will actually only prove smoothness of the densities $P(t, x, y)$ in y. Joint regularity in (x, y) can be obtained by using these techniques in conjunction with the stability theorems of section 3.2 (see 'Malliavin calculus for processes with jumps', K. Bichteler and J. Jacod, preprint, 1984).

1

Background material

The purpose of this chapter is to provide a brief introduction to the theories of Gaussian measures on Banach spaces and stochastic integration with respect to Brownian motion and also to collect together some results which will be needed later. The notation introduced here will remain in force throughout the book.

1.1 Abstract Wiener spaces

Definition 1.1 A Borel measure γ on a real Banach space E is said to be *Gaussian* if the projection $e(\gamma)$ of γ onto $\mathbf{R}$ is a normal distribution for every e in E^*.

We will now describe a procedure, due to L. Gross, for constructing a Gaussian measure on an abstract Banach space which generalizes Norbert Wiener's original construction of the Wiener measure on the space of paths. Suppose that H is a separable Hilbert space with inner product $\langle , \rangle$. For every surjective linear map T from H to a Euclidean space $\mathbf{R}^n$ define $\langle , \rangle_T$ to be the inner product on $\mathbf{R}^n$ such that T is an isometry from $((\ker T)^{\perp}, \langle \, \rangle)$ to $(\mathbf{R}^n, \langle , \rangle_T)$. Define further a Gaussian measure μ_T on $\mathbf{R}^n$ by

$$\mu_T(B) = (2\pi)^{-n/2} \int_B \mathrm{e}^{-1/2\langle x, x\rangle_T} \, \mathrm{d}x \tag{1.1}$$

for Borel sets B in $\mathbf{R}^n$, where $\mathrm{d}x$ is the Lebesgue measure on $\mathbf{R}^n$ generated by the inner product $\langle , \rangle_T$. Let $\mathcal{R}_H$ denote the ring of cylinder sets in H, i.e. the collection of subsets of H of the form $T^{-1}(B)$ where $T \in L(H, \mathbf{R}^n)$ for some $n \in \mathbf{N}$ and B is a Borel set in $\mathbf{R}^n$. Finally, define a function μ on $\mathcal{R}_H$ by

$$\mu(T^{-1}(B)) = \mu_T(B).$$

It can be shown that μ is a well defined finitely additive 'measure' on $\mathscr{R}_H$. Unfortunately it turns out that if H is infinite dimensional, then μ is not countably additive; hence it cannot be extended to a measure on the completion of $\mathscr{R}_H$. However, one can construct a measure from μ in the following way. Suppose that E is a separable Banach space and let i be a continuous linear injection from H into E with dense range. Then $i(\mu)$ defines a positive additive function on the ring $\mathscr{R}_E$ of cylinder sets in E. Gross has shown that if the norm on E is measurable with respect to H, then $i(\mu)$ is actually countably additive on $\mathscr{R}_E$ (see Gross[12] for the definition of measurable norm and a proof of this result). In this case the cylinder set measure $i(\mu)$ extends to give a measure γ on the σ-field $\bar{\mathscr{R}}_E$ generated by $\mathscr{R}_E$. Since E is separable, $\bar{\mathscr{R}}_E$ is the Borel field of E, hence γ is a Borel measure on E. The triple (i, H, E) is known as an *abstract Wiener space*. It can be shown that every separable Hilbert space is the H of an abstract Wiener space and that every separable Banach space E contains a densely embedded Hilbert space H such that (i, H, E) is an abstract Wiener space.

For the remainder of this section let (i, H, E) be an abstract Wiener space where H has inner product $\langle , \rangle$ and norm $\|\cdot\|$ and let γ denote the corresponding Borel measure on E. For notational convenience we will identify E^* and H with their images in E under the inclusions

$$E^* \xrightarrow{i^*} H^* \cong H \xrightarrow{i} E.$$

Theorem 1.2 *γ is a Gaussian measure. In fact for every e in E^*, $e(\gamma)$ is the normal measure $N(0, \|e\|^2)$.*

This follows directly from (1.1).

Theorem 1.3 *If H is infinite dimensional, then $\gamma(H) = 0$.*

Proof Let $\{e_i\}_{i=1}^{\infty}$ be an orthonormal basis of H which is contained inside E^*. Note that

$$H = \left\{x \in E : \sum_{i=0}^{\infty} e_i(x)^2 < \infty\right\}.$$

Since (1.1) implies that $\{e_i(x)\}_{i=1}^{\infty}$ is a sequence of independent $N(0, 1)$ random variables under γ, it follows from Kolmogorov's 0–1 law that $\gamma(H)$ is either 0 or 1. Let $\varepsilon > 0$ and $M > N$. Then

$$\gamma\left\{x : \sum_{i=N}^{M} e_i(x)^2 > \varepsilon\right\} \geq \gamma\{x : |e_N(x)| > \sqrt{\varepsilon}\} = c$$

where c is a positive constant independent of M and N. This implies that the sequence

$$\left\{\sum_{i=1}^{M} e_i(x)^2\right\}_{M\geq 1}$$

is divergent in probability. Thus $\gamma(H)=0$. ■

The following result is an integration by parts theorem for abstract Wiener space. A proof can be found in Elworthy[7].

Theorem 1.4 (Divergence theorem) *Suppose that f and g are C^1 maps from E into $\mathbf{R}$ and E into E^* respectively. Then*

$$\int_E Df(x)g(x)\,\mathrm{d}\gamma(x) = \int_E f(x)[\mathrm{Div}\,g](x)\,\mathrm{d}\gamma(x)$$

provided that either side exists, where Div g *denotes the divergence of* g *with respect to* γ, *defined by*

$$[\mathrm{Div}\,g](x) = g(x)x - \mathrm{Trace}_H\,Dg(x). \tag{1.2}$$

The abstract Wiener space with which we shall be primarily concerned is the following. For a fixed positive integer n let C_0 $(=C_0(\mathbf{R}^n))$ denote the Banach space of paths from $[0,1]$ into $\mathbf{R}^n$ with initial point 0, equipped with the supremum norm. Let $L_0^{2,1}$ $(=L_0^{2,1}(\mathbf{R}^n))$ denote the Hilbert subspace of C_0 consisting of absolutely continuous paths with square integrable derivatives with inner product $\langle\,,\rangle$ defined by

$$\langle \sigma, \tau\rangle = \int_0^1 \langle \sigma'(t), \tau'(t)\rangle_{\mathbf{R}^n}\,\mathrm{d}t; \qquad \sigma, \tau \in L_0^{2,1}$$

where $\langle\,,\rangle_{\mathbf{R}^n}$ denotes the standard inner product on $\mathbf{R}^n$. Finally let i be the inclusion map from $L_0^{2,1}$ into C_0. Then one can show that $(i, L_0^{2,1}, C_0)$ is an abstract Wiener space. The corresponding Gaussian measure γ on C_0 (which we refer to in the sequel as Wiener measure) is actually the n-fold product of the measure originally constructed by Wiener.

Suppose that $(\Omega, \mathscr{F}, P)$ is a probability triple and $w:\Omega\to C_0$ is a random variable with distribution γ. Then it follows from the definition of γ and (1.1) that w has the following properties

(i) For every $0\leq s<t\leq 1$ the distribution of $w_t - w_s$ is the n-fold product of $N(0, t-s)$ distributions.

(ii) For every $0\leq s<t\leq 1$ the random variable $w_t - w_s$ is independent of $\{w_v : 0\leq v\leq s\}$.

Obviously w also satisfies

(iii) $w_0 = 0$ a.s.

(iv) The map from $[0,1]$ to $\mathbf{R}^n$ defined by $t\to w_t$ is continuous a.s.

A random variable w with the above properties (i)–(iv) is said to be (*normalized*) *n-dimensional Brownian motion* (with respect to P).

For every $t \in [0, 1]$, let $\mathcal{F}_t$ denote the σ-field generated by $\{w_s : s \leq t\}$. We will say that a process $\theta : \Omega \times [0, 1] \to \mathbf{R}^d$ is *adapted* (to $\{\mathcal{F}_t\}$) if θ_t is measurable with respect to $\mathcal{F}_t$, for all $0 \leq t \leq 1$.

1.2 Stochastic integration

Let w denote an n-dimensional Brownian motion defined on a probability space $(\Omega, \mathcal{F}, P)$ and let E denote expectation with respect to P. For each $0 \leq t \leq 1$ let $\mathcal{F}_t$ be the σ-field generated by w up to time t, i.e.

$$\mathcal{F}_t = \sigma(w_s : 0 \leq s \leq t).$$

The following definition introduces the class of stochastically integrable processes.

Definition 1.5 Let $\mathcal{M}_1$ denote the class of maps f from $\Omega \times [0, 1] \to \mathbf{M}^{n,d}$, the set of $n \times d$ real-valued matrices, such that
(i) f is jointly measurable with respect to $\mathcal{F} \times \beta$ (where β denotes the Borel field on $[0, 1]$).
(ii) $f(\cdot, t)$ is $\mathcal{F}_t$-measurable for each $0 < t < 1$.
(iii) $\int_0^1 \|f(\cdot, t)\|^2 \, dt < \infty$ a.s., where $\| \; \|$ denotes the Hilbert–Schmidt norm on $\mathbf{M}^{n,d}$.

Definition 1.6 Let $\mathcal{M}_2 = \{f \in \mathcal{M}_1 :$ there exists a (non-random) partition $0 = t_0 < t_1 < \ldots < t_n = 1$ and a (non-random) constant C such that
(i) $f(\omega, t) = f(\omega, t_i)$ if $t_i \leq t < t_{i+1}$.
(ii) $\|f(\omega, t)\| \leq C$ for all $(\omega, t) \in \Omega \times [0, 1]\}$.

For $f \in \mathcal{M}_2$ and any $T \in [0, 1]$ we define

$$\int_0^T f(s) \, dw_s = \sum_{i=1}^{m-1} f(t_i)[w_{t_{i+1}} - w_{t_i}] + f(t_m)[w_T - w_{t_m}] \tag{1.3}$$

where $t_1, t_2, \ldots, t_m$ are as in Definition 1.6 and t_m is the largest stepping time of f less than T. In order to extend this definition to functions in $\mathcal{M}_1$ we need the following results (see Gikhman and Skorohod[11] for proofs).

Lemma 1.7 *Suppose that $f \in \mathcal{M}_2$ and $T \in [0, 1]$. Then for all positive numbers C and D the stochastic integral defined in* (1.3) *has the property*

$$P\left(\left\|\int_0^T f(s) \, dw_s\right\| > D\right) \leq C/D^2 + P\left(\int_0^T \|f(s)\|^2 \, ds > C\right) \tag{1.4}$$

where the norm on the left-hand side is the ℓ^2-norm on $\mathbf{R}^d$.

Lemma 1.8 *For each $f \in \mathcal{M}_1$ there exists a sequence $\{f_n\}_{n\geq 1}$ contained in $\mathcal{M}_2$ such that*

$$\int_0^1 \|f(s) - f_n(s)\|^2 \, ds \to 0 \quad a.s. \tag{1.5}$$

Suppose that $f \in \mathcal{M}_1$ and let $\{f_n\}_{n\geq 1}$ be a sequence as in Lemma 1.8. Then (1.5) and (1.4) imply that for each $T \in [0, 1]$, $\int_0^T f_n(s)\, dw_s$ is a Cauchy sequence in probability. The (*Itô*) *stochastic integral* is defined by

$$\int_0^T f(s)\, dw_s = \lim_{n\to\infty} (\text{probability}) \int_0^T f_n(s)\, dw_s.$$

Note that (1.5) and (1.4) imply that this limit is independent of the choice of the sequence $\{f_n\}_{n\geq 1}$. The reader is referred to Gikhman and Skorohod[11] for the proof of the following two results.

Theorem 1.9 *The stochastic integral has the following properties. Let $T \in [0, 1]$ and $f \in \mathcal{M}_1$. Then*

(i) (Additivity). *For any $0 \leq T_1 \leq T \leq 1$*

$$\int_0^T f(s)\, dw_s = \int_0^{T_1} f(s)\, dw_s + \int_{T_1}^T f(s)\, dw_s \quad a.s.$$

(ii) (Linearity). *For every $g \in \mathcal{M}_1$ and random variable K*

$$\int_0^T [f(s) + Kg(s)]\, dw_s = \int_0^T f(s)\, dw_s + K\int_0^T g(s)\, dw_s \quad a.s.$$

(iii) *If $\int_0^T E\,\|f(s)\|^2\, ds < \infty$, then*

$$E\int_0^T f(s)\, dw_s = 0 \quad \text{and} \quad E\left\|\int_0^T f(s)\, dw_s\right\|^2 = \int_0^T E\,\|f(s)\|^2\, ds.$$

(iv) *If $E\int_0^T \|f(s)\|^{2m}\, ds < \infty$ for some $m > 1$, then*

$$E\left[\sup_{0\leq s\leq T}\left\|\int_0^s f(u)\, dw_u\right\|^{2m}\right] \leq m^{2m}\left(\frac{2m-1}{m-1}\right)^m T^{m-1}\int_0^T E\,\|f(s)\|^{2m}\, ds$$

(v) *There exists an a.s. continuous version of the process*

$$\left\{\int_0^t f(s)\, dw_s : t \in [0, 1]\right\}.$$

(vi) (Itô's lemma). *Suppose that $g : \Omega \times [0, 1] \to \mathbf{R}^d$ satisfies* (i) *and* (ii) *of Definition* 1.5 *and in addition g is Lebesgue integrable on* $[0, 1]$ *a.s. Let $x \in \mathbf{R}^d$ and define*

$$\xi_t = x + \int_0^t f(s)\, dw_s + \int_0^t g(s)\, ds, \qquad t \in [0, 1].$$

Suppose further that F is a C^2 function from $\mathbf{R}^d$ to a Euclidean space $\mathbf{R}^m$. Then for each $t \in [0, 1]$ $\eta_t = F(\xi_t)$ satisfies

$$\eta_t = F(x) + \int_0^t DF(\xi_s)f(s)\,\mathrm{d}w_s + \int_0^t \Big\{DF(\xi_s)g(s) + \frac{1}{2}\sum_{i=1}^n D^2F(\xi_s)(f_i(s), f_i(s))\Big\}\,\mathrm{d}s; \quad a.s.$$

where $f_i(s)$ is the ith *column of the matrix $f(s)$, for $i = 1, \ldots, n$.*

Definition 1.10 Suppose that $T \in [0, 1]$, f, g and ξ are as in Theorem 1.9(vi), $m \in \mathbf{N}$ and G is a C^1 function from $\mathbf{R}^d$ to $\mathbf{M}^{n,m}$. Then we define the *Stratonovich integral*, $\int_0^T G(\xi_s)\,\mathrm{d}\circ w_s$ as

$$\int_0^T G(\xi_s)\,\mathrm{d}\circ w_s = \int_0^T G(\xi_s)\,\mathrm{d}w_s + \frac{1}{2}\sum_{i=1}^n \int_0^T DG_i(\xi_s)f_i(\xi_s)\,\mathrm{d}s.$$

The Stratonovich integral has the advantage that it transforms under composition with a C^2 map in the same way as a classical Riemann–Stieltjes integral (i.e. by the chain rule). Thus Stratonovich stochastic calculus takes the same form as classical integral calculus.

Theorem 1.11 Suppose that $x \in \mathbf{R}^d$ and that $A : \mathbf{R}^d \to \mathbf{M}^{n,d}$ and $B : \mathbf{R}^d \to \mathbf{R}^d$ are maps satisfying the following conditions:

(i) *There exists a constant $D > 0$ such that for all x in $\mathbf{R}^d$*

$$\|A(x)\|^2 \le D(1 + \|x\|^2)$$
$$\|B(x)\|^2 \le D(1 + \|x\|^2)$$

(ii) *A and B are Lipschitz, i.e. there exists a constant L such that for all x and y*

$$\|A(x) - A(y)\| \le L\,\|x - y\|$$
$$\|B(x) - B(y)\| \le L\,\|x - y\|$$

Then there exists an adapted process $\xi : \Omega \times [0, 1] \to \mathbf{R}^d$ with a.s. continuous sample paths such that

$$\xi_t = x + \int_0^t A(\xi_s)\,\mathrm{d}w_s + \int_0^t B(\xi_s)\,\mathrm{d}s, \qquad t \in [0, 1] \quad a.s. \tag{1.6}$$

Furthermore the process ξ is unique up to stochastic equivalence.

Note that if A and B are as in Theorem 1.11, A is C^1 and for each $1 \le i \le n$ the function $x \to DA_i(x)A_i(x)$ satisfies the conditions on B in

Theorem 1.11, then one can define the Stratonovich equation

$$\xi_t = x + \int_0^t A(\xi_s)\,\mathrm{d}\circ w_s + \int_0^t B(\xi_s)\,\mathrm{d}s, \quad t \in [0, 1].$$

This can be rewritten in Itô form as

$$\xi_t = x + \int_0^t A(\xi_s)\,\mathrm{d}w_s + \int_0^t \left[\frac{1}{2}\sum_{i=1}^n DA_i(\xi_s)A_i(\xi_s) + B(\xi_s)\right]\mathrm{d}s,$$
$$t \in [0, 1].$$

1.3 Preliminary results

Let $\mathscr{C}$ denote the class of test functions on $\mathbf{R}^d$ (i.e. real-valued C^∞ functions defined on $\mathbf{R}^d$ with compact support) and suppose that ν is a finite Borel measure on $\mathbf{R}^d$. Lemmas 1.12 and 1.14 below are the basis of the regularity results which we will prove in chapters 2–4.

Lemma 1.12 *Suppose that for every vector y in $\mathbf{R}^d$ there exists a constant C such that the following inequality holds for all functions φ in $\mathscr{C}$*

$$\left|\int_{\mathbf{R}^d} D\varphi(x)y\,\mathrm{d}\nu(x)\right| \le C\,\|\varphi\|_\infty \tag{1.7}$$

Then ν is absolutely continuous with respect to the Lebesgue measure on $\mathbf{R}^d$ (see Malliavin[19]).

Proof See Malliavin[19]. ■

In Bell[3] we prove the following† sharp form of the last result.

Theorem 1.13 *The following conditions on ν are equivalent.*
(i) *There exists an open set U in $\mathbf{R}^d$ of full ν-measure and measurable functions $X_1, X_2, \ldots, X_d$, continuous on U such that the relations*

$$\int_{\mathbf{R}^d} D\varphi(x)e_i\,\mathrm{d}\nu(x) = \int_{\mathbf{R}^d} \varphi(x)X_i(x)\,\mathrm{d}\nu(x)$$

hold for all φ in $\mathscr{C}$ with support $\varphi \subset U$ and $i = 1, \ldots, d$, where $\{e_1, e_2, \ldots, e_d\}$ is an orthonormal basis of $\mathbf{R}^d$.

† This will not actually be used in the sequel.

(ii) *The measure ν is absolutely continuous with respect to Lebesgue measure λ on $\mathbf{R}^d$, and the density function F of ν is C^1 on an open set V of full ν-measure.*

Lemma 1.14 *Suppose that for every $b \in \mathbf{N}$ and unit vectors $y_1, y_2, \ldots, y_b \in \mathbf{R}^d$ there exists a constant C_b such that for all $\varphi \in \mathscr{C}$*

$$\left|\int_{\mathbf{R}^d} D^b\varphi(x)(y_1, \ldots, y_b)\,\mathrm{d}\nu(x)\right| \leq C_b\,\|\varphi\|_\infty \tag{1.8}$$

Then the density F of ν is a C^∞ function. Furthermore for every $n > d$, $\|D^{n-d-1}F\|_\infty \leq A(n, d)C^{(n)}$, where $A(n, d)$ is a constant depending only on n and d, and $C^{(n)} \equiv \max\{C_1, \ldots, C_n\}$.

Proof See Stroock[27]. ■

Theorem 1.15 (Girsanov theorem). *As before, let $w : \Omega \times [0, 1] \to \mathbf{R}^n$ denote n-dimensional Brownian motion and suppose that $\varphi : \Omega \times [0, 1] \to \mathbf{R}^n$ is a bounded process adapted to $\{\mathscr{F}_t\}$. Define*

$$\tilde{w}_t = w_t - \int_0^t \varphi_s\,\mathrm{d}s, \qquad t \in [0, 1]$$

and a probability measure $\tilde{P}$ on Ω by

$$\frac{\mathrm{d}\tilde{P}}{\mathrm{d}P} = \exp\left\{\sum_{i=1}^n \int_0^1 \varphi_i(s)\,\mathrm{d}w_i(s) - \frac{1}{2}\int_0^1 \|\varphi_s\|^2\,\mathrm{d}s\right\}.$$

Then $\tilde{w}$ is normalized Brownian motion with respect to $\tilde{P}$.

Proof See Friedman[9]. ■

Theorem 1.16 (Wiener decomposition theorem). *Let γ denote the Wiener measure on $C_0(\mathbf{R}^n)$ and $w = (w_1, \ldots, w_n)$, n-dimensional Brownian motion. Define $\bar{Z}_0 \equiv \mathbf{R}$ and for each $m \in \mathbf{N}$ define $\Delta_m \equiv \{(t_1, \ldots, t_m) \in \mathbf{R}^m : 0 \leq t_1 < \ldots < t_m \leq 1\}$ and*

$$\bar{Z}_m \equiv \operatorname{span}\left\{\int_0^1 \cdots \int_0^{t_3}\int_0^{t_2} f(t_1, \ldots, t_m)\,\mathrm{d}w_{j_1}(t_1)\,\mathrm{d}w_{j_2}(t_2)\ldots\mathrm{d}w_{j_m}(t_m) : \right.$$

$$\left. f \in L^2(\Delta_m), \qquad 1 \leq j_1, j_2, \ldots, j_m \leq n\right\}, \qquad m \in \mathbf{N}.$$

Then $\{\bar{Z}_m : m = 0, 1, \ldots\}$ are orthogonal subspaces of $L^2(\gamma)$. Furthermore

$$L^2(\gamma) = \bigoplus_{m=0}^\infty \bar{Z}_m.$$

Proof See Stroock[26]. ■

Theorem 1.17 (Gronwall's lemma). Let h and k be real-valued Lebesgue integrable functions on $[0, T]$ *such that for some* $D>0$

$$h_t \leq k_t + D\int_0^t h_s \, \mathrm{d}s, \qquad t \in [0, T].$$

Then

$$h_t \leq k_t + D\int_0^t \mathrm{e}^{D(t-s)} k_s \, \mathrm{d}s \qquad \text{for almost all} \quad t \in [0, T].$$

Proof See Elworthy[8]. ■

Notation Let (E, μ) be a measure space and let T be a measurable map from E to a Banach space F. Then $T(\mu)$ will denote the measure induced on F by T, defined by $T(\mu)(B) = \mu(T^{-1}B)$, for any Borel set $B \subset F$.

2

The functional analytic approach

2.1 Symmetric diffusion operators

The work of Stroock[26], on which this chapter is based, is a functional analytic formulation of the theory outlined by Malliavin in his paper[19] of 1976. Here the integration by parts operation required in order to derive regularity results for the measures induced by stochastic integral maps is performed via the application of a symmetric linear operator and an associated bilinear form; both maps being defined on a dense subspace of the space of L^2 Wiener functionals. These operations, which comprise the basis of Malliavin's calculus, are manipulated as indicated in our introduction to yield the required regularity results.

Stroock axiomatizes the essential properties of Malliavin's operator and defines such an object with respect to a general Gaussian measure. The result is the following.

Definition 2.1 A densely defined linear operator $\mathscr{L}$ on $L^2(\gamma)$ with domain $\mathscr{D}(\mathscr{L})$ is a *symmetric diffusion operator* (SDO) if it has the following properties:
(i) $\mathscr{L}$ is self-adjoint on $\mathscr{D}(\mathscr{L})$.
(ii) $1 \in \mathscr{D}(\mathscr{L})$ and $\mathscr{L}1 = 0$.
(iii) There is a linear subspace $\mathscr{D} \subseteq \{\phi \in \mathscr{D}(\mathscr{L}) \cap L^4(\gamma)$: $\mathscr{L}\phi \in L^4(\gamma)$ and $\phi^2 \in \mathscr{D}(\mathscr{L})\}$ such that graph $(\mathscr{L}/\mathscr{D})$ is dense in graph $(\mathscr{L})$.
(iv) Define $\langle \phi, \psi \rangle = \mathscr{L}(\phi\psi) - \phi\mathscr{L}\psi - \psi\mathscr{L}\phi$ for $(\phi, \psi) \in \mathscr{D} \times \mathscr{D}$. Then $\langle \cdot, \cdot \rangle : \mathscr{D} \times \mathscr{D} \to L^2(\gamma)$ is a non-negative bilinear form.
(v) If $\phi = (\phi_1, \ldots, \phi_n) \in \mathscr{D}^n$ and $F \in C_b^2(\mathbf{R}^n)$, the set of bounded C^2 functions on $\mathbf{R}^n$ with bounded first and second derivatives; then $F \circ \phi \in \mathscr{D}(\mathscr{L})$ and

$$\mathscr{L}(F \circ \phi) = \frac{1}{2} \sum_{i,j=1}^{n} \langle \phi_i, \phi_j \rangle \frac{\partial^2 F}{\partial x_i\, \partial x_j} \circ \phi + \sum_{i=1}^{n} \mathscr{L}\phi_i \frac{\partial F}{\partial x_i} \circ \phi.$$

(vi) $\mathscr{L}$ admits an extension $\mathscr{L}_1$ to a closed linear operator on $L^1(\gamma)$ with domain $\mathscr{D}(\mathscr{L}_1)$ such that $\mathscr{D}(\mathscr{L}) = \{\phi \in \mathscr{D}(\mathscr{L}_1) \cap L^2(\gamma) : \mathscr{L}_1\phi \in \mathscr{L}^2(\gamma)\}$.

Recall from chapter 1 that a Gaussian measure on a Banach space determines a densely embedded Hilbert space H. An instructive way to view $\mathscr{L}$ is as a closed extension in $L^2(\gamma)$ of the differential operator $\tilde{\mathscr{L}}g(x) = 1/2(\text{Trace } D^2g(x) - Dg(x)x)$ where the latter is defined on a class of suitably smooth functions on H which contains $\mathscr{D}$. When E is finite dimensional, then $H = E$, and $\tilde{\mathscr{L}}$ is known as the Ornstein–Uhlenbeck operator. Lemma 2.2 illustrates a natural relationship between $\tilde{\mathscr{L}}$ and γ (in the one-dimensional case). In general the properties of $\tilde{\mathscr{L}}$ motivate much of Definition 2.1 and in particular parts (iv) and (v). Part (vi) is used in order to check that the domain of $\mathscr{L}$ is closed under algebraic operations and compositions with regular functions.

As some insight is gained into the structure of the operator for Wiener space introduced in section 2.2 by considering the situation where $E = \mathbf{R}$ and $d\gamma(x) = 1/\sqrt{2\pi}\, e^{-x^2/2}\, dx$, we will first define and establish the properties of an SDO in this simpler case.

Let $\{H_n : n \geq 0\}$ be the orthonormal basis of $L^2(\gamma)$ consisting of Hermite polynomials, i.e.

$$H_n(x) = \frac{(-1)^n}{\sqrt{n!}} e^{x^2/2} D^n e^{-x^2/2}; \qquad n \geq 0. \tag{2.1}$$

Define $\tilde{\mathscr{L}} = 1/2(D^2 - xD)$ on C^∞ functions on $\mathbf{R}$. Then the Hermite polynomials are eigenvectors of $\tilde{\mathscr{L}}$. In fact one has

Lemma 2.2 $\tilde{\mathscr{L}}H_n = -nH_n/2$ *for each* $n \geq 0$. (2.2)

Proof For $f \in C^\infty$, define $R(f) = e^{x^2/2}D(e^{-x^2/2}f)$. Then

$$RD - DR = I. \tag{2.3}$$

Multiplying on the left by R and using (2.3) gives

$$R = R^2D - RDR = R^2D - (DR + I)R,$$

thus

$$R^2D - DR^2 = 2R.$$

Continuing in this way one obtains

$$R^nD - DR^n = nR^{n-1}, \qquad n \geq 1.$$

Applying this to 1 and using $H_n = R^n(1)$ gives $-DH_n = nH_{n-1}$. In conjunction with the relation $(D - X)H_n = H_{n+1}$ this yields (2.2) for $n \geq 1$. Finally, note that (2.2) is true for $n = 0$. ■

Since the Hermite polynomials form an orthonormal basis of $L^2(\gamma)$ if one defines

$$\mathscr{D}(\mathscr{L}) = \left\{\phi \in L^2(\gamma): \sum_{n=1}^{\infty} n^2(\phi, H_n)^2 < \infty\right\}$$

and

$$\mathscr{L}\phi = \sum_{n=1}^{\infty} -n/2(\phi, H_n)$$

where $(\cdot, \cdot)$ is the inner product on $L^2(\gamma)$, then it follows that $\mathscr{L}$ is closed on $L^2(\gamma)$. If we let $\mathscr{D} = \text{span}\{H_n : n \geq 0\}$, then by Lemma 2.2 $\mathscr{L}$ extends $\tilde{\mathscr{L}}$ where the latter is considered as an operator on $\mathscr{D}$.

Theorem 2.3 *$\mathscr{L}$ is an SDO with respect to γ.*

Proof It is clear that $\mathscr{L}$ satisfies parts (i) and (ii) of Definition 2.1. Part (iii) follows from the fact that for each N, $\text{span}\{H_n : n \leq N\} = \text{span}\{x^n : n \leq N\}$.

Suppose that $\phi \in L^2(\gamma)$ is such that the sequence $\{(\phi, H_n)\}_1^{\infty}$ is rapidly decreasing. If we let

$$\phi_N = \sum_{n=1}^{N} (\phi, H_n)H_n$$

then

$$\mathscr{L}\phi_N = 1/2(D^2 - xD)\phi_N.$$

The right-hand side converges in $L^2(\gamma)$ to $1/2(D^2 - xD)\phi$, and since $\mathscr{L}$ is closed it follows that $\phi \in \mathscr{D}(\mathscr{L})$ and satisfies

$$\mathscr{L}\phi = 1/2(D^2 - xD)\phi. \tag{2.4}$$

Let $C_\uparrow^2(\mathbf{R}) = \{\phi \in C^2(\mathbf{R}) : \phi, \phi', \phi'' \text{ are slowly increasing}\}$. If $\psi \in C_\uparrow^2(\mathbf{R})$, then by approximating ψ by functions as in (2.4) one sees that $\psi \in \mathscr{D}(\mathscr{L})$ and also satisfies (2.4). Now if $F \in C_\uparrow^2(\mathbf{R})$ and $(\phi_1, \ldots, \phi_n) \in \mathscr{D}^n$, then $F \circ \phi \in C_\uparrow^2(\mathbf{R})$ and from (2.4) we have

$$\begin{aligned}\mathscr{L}(F \circ \phi) &= 1/2(D^2 - xD)(F \circ \phi)\\ &= \frac{1}{2}\sum_{i,j=1}^{n} \phi_i'\phi_j' \frac{\partial^2 F}{\partial x_i\, \partial x_j} \circ \phi + \sum_{i=1}^{n} \frac{1}{2}(\phi_i'' - x\phi_i')\frac{\partial F}{\partial x_i} \circ \phi\\ &= \frac{1}{2}\sum_{i,j=1}^{n} \phi_i'\phi_j' \frac{\partial^2 F}{\partial x_i\, \partial x_j} \circ \phi + \sum_{i=1}^{n} \mathscr{L}\phi_i \frac{\partial F}{\partial x_i} \circ \phi.\end{aligned} \tag{2.5}$$

This shows that $\mathscr{L}$ satisfies (v). Furthermore taking $F(x, y) = x \cdot y$ in (2.5) shows that for $\phi, \psi \in \mathscr{D}$, $\langle \phi, \psi \rangle = \phi' \psi'$ and so $\mathscr{L}$ also satisfies (iv).

Part (vi) can be checked by considering the Ornstein–Uhlenbeck process, defined by the equation

$$\eta_t = X + w_t - \frac{1}{2} \int_0^t \eta_s \, ds, \qquad t \geq 0. \tag{2.6}$$

Here X is a random variable with distribution γ and $\{w_t\}_{t\geq 0}$ normalized Brownian motion independent of X. Solving (2.6) for η gives

$$\eta_t = e^{-t/2}\left(X + \int_0^t e^{s/2} \, dw_s\right), \qquad t \geq 0. \tag{2.7}$$

Let $\{T_t\}_{t\geq 0}$ be the semi-group defined by $(T_t\phi)(X) = E[\phi(\eta_t)/X]$ acting on $C_b(\mathbf{R})$, the set of bounded functions on $\mathbf{R}$. Equation (2.7) implies that η is a stationary process with invariant measure γ. So for every $p \geq 1$

$$\begin{aligned} \|T_t\phi\|_{L^p}^p = E[|E[\phi(\eta_t)/X]|^p] &\leq E[|\phi(\eta_t)|^p] \\ &= \|\phi\|_{L^p}^p. \end{aligned}$$

It follows from the Hahn–Banach theorem that $\{T_t\}_{t\geq 0}$ extends to a contraction semi-group $\{T_t^p\}_{t\geq 0}$ on $L^p(\gamma)$. Let $\mathscr{L}_p$ be the generator; then $\mathscr{L}_p$ is a closed operator with respect to the L^p-norm for every $p \geq 1$. The application of Itô's lemma to equation (2.6) shows that for $\phi \in C_\uparrow^2(\mathbf{R})$

$$\lim_{t\downarrow 0} \left(\frac{T_t^2\phi - \phi}{t}\right)(X) \stackrel{L_2}{=} \tfrac{1}{2}(D^2\phi(X) - XD\phi(X)) = \mathscr{L}\phi(X).$$

Thus $\mathscr{L} = \mathscr{L}_2$ on $C_\uparrow^2(\mathbf{R})$, and since graph$(\mathscr{L}/C_\uparrow^2(\mathbf{R}))$ is dense in graph $\mathscr{L}$ and $\mathscr{L}_2$ is closed, it follows that $\mathscr{L} \subseteq \mathscr{L}_2$. However, equation (2.7) also implies that T_t for each t and hence $\mathscr{L}$ are self-adjoint operators on $L^2(\gamma)$ and since this is also true of $\mathscr{L}$ one concludes that $\mathscr{L} = \mathscr{L}_2$.† Thus $\mathscr{L}_1$ is a closed extension of $\mathscr{L}$ on $L^1(\gamma)$. Suppose that $\phi \in \mathscr{D}(\mathscr{L}_1) \cap L^2(\gamma)$ and $\mathscr{L}_1\phi \in L^2(\gamma)$. Then by the usual semi-group relation we have

$$T_t^2\phi - \phi = T_t^1\phi - \phi = \int_0^t T_s^1\mathscr{L}_1\phi \, ds = \int_0^t T_s^2\mathscr{L}_1\phi \, ds. \tag{2.8}$$

Since T_t^2 is pointwise continuous on $L^2(\gamma)$, (2.8) shows that $\lim_{t\downarrow 0}(T_t^2\phi - \phi)/t$ exists in L^2, i.e. $\phi \in \mathscr{D}(\mathscr{L})$ and (vi) is satisfied. This completes the proof of Theorem 2.3. ■

† This follows from the fact that a self-adjoint operator admits no strict self-adjoint extension.

2.2 A symmetric diffusion operator for Wiener space

Throughout this section, γ will denote the Wiener measure on the space C_0 of paths from $[0, 1]$ into $\mathbf{R}^n$ with initial point 0; where C_0 is endowed with the supremum norm and the Borel σ-algebra. In order to simplify notation we shall, following Stroock's paper[26], describe the construction of an SDO $\mathscr{L}$ for γ in the case $n=1$. No essential difficulties are encountered when $n>1$, and the modifications required when dealing with the multi-dimensional Wiener measure will be discussed at the end of the section. Suppose then that $n=1$. The form of $\mathscr{L}$ is motivated as follows.

Consider the finite dimensional measure space $(\mathbf{R}^n, \beta^n, \gamma^{(n)})$ where β is the Borel field on $\mathbf{R}$ and $\gamma^{(n)}$ is the product of standard normal measures $\{\gamma_i\}_{i=1}^n$ on β. Since the vectors $\{H_\alpha(x)=H_{\alpha_1}(x_1)\ldots H_{\alpha_n}(x_n)\}$ form an orthonormal basis of $L^2(\gamma^{(n)})$, a straightforward extension of Theorem 2.3 shows that the operator $\mathscr{L}^{(n)}$ defined on $L^2(\gamma^{(n)})$ by the action $\mathscr{L}^{(n)}H_\alpha=-(|\alpha|/2)H_\alpha$ on each H_α, where $|\alpha|=\alpha_1+\ldots+\alpha_n$, is an SDO for $\gamma^{(n)}$. Now let Γ denote the measure $\bigotimes_{i=1}^\infty \gamma_i$ on the infinite product space $(\mathbf{R}^\infty, \beta^\infty)$. The Fock space $L^2(\Gamma)$ has an orthonormal basis $\{H_\alpha(x)=H_{\alpha_1}(x_1)\ldots H_{\alpha_n}(x_n): n\geq 0,\ 0\leq \alpha_1, \ldots, \alpha_n<\infty\}$ and a natural object to consider as an SDO for Γ is the operator $\mathscr{L}_F$ obtained by setting

$$\mathscr{D}(\mathscr{L}_F)=\left\{\phi=\sum a_\alpha H_\alpha \in L^2(\Gamma): \sum |\alpha|^2 a_\alpha^2<\infty\right\}$$

and for $\phi=\sum a_\alpha H_\alpha \in \mathscr{D}(\mathscr{L}_F)$, defining

$$\mathscr{L}_F\phi=\sum -\frac{|\alpha|}{2} a_\alpha H_\alpha.$$

$\mathscr{L}_F$ transfers to an operator $\mathscr{L}$ on $L^2(\gamma)$ by identifying $L^2(\gamma)$ and $L^2(\Gamma)$ in the following way: suppose that $\{f_n^\infty\}_{n=1}$ is a fixed orthonormal basis of $L^2[0, 1]$ and for $w\in(C_0, \gamma)$ let

$$\vec{w}=\left(\int_0^1 f_1\,dw, \int_0^1 f_2\,dw, \ldots\right)$$

where these are all Itô integrals. Then $\vec{\ }$ is a measure-preserving map between (C_0, γ) and $(\mathbf{R}^\infty, \beta)$. In fact if one defines $T: L^2(\Gamma)\to L^2(\gamma)$ by $TF(w)=F(\vec{w})$ for $F\in L^2(\Gamma)$, then it turns out that T is an isometric isomorphism between the L^2 spaces (cf. Stroock[26], p. 408). Define $\mathscr{L}$ to be the operator on $L^2(\gamma)$ with domain $T(\mathscr{D}(\mathscr{L}_F))$ by the relation $\mathscr{L}=T\circ\mathscr{L}_F\circ T^{-1}$.

A more satisfactory description of $\mathscr{L}$ can be obtained by invoking Wiener's theory of homogeneous chaos. Define $Z_0=\mathbf{R}$ and for each

$n \geq 1$, $Z_n = \text{span}\{H_\alpha(\vec{w}) : |\alpha| = n\}$. It can be shown (cf. Stroock[26], p. 408), that $\bar{Z}_n$ in $L^2(\gamma)$ is the space consisting of stochastic integrals of the form

$$\int_0^1 \int_0^{t_n} \cdots \int_0^{t_2} f(t_1, \ldots, t_n)\, \mathrm{d}w(t_1) \ldots \mathrm{d}w(t_n)$$

where $f : [0, 1]^n \to \mathbf{R}$ is deterministic and

$$\int_0^1 \int_0^{t_n} \cdots \int_0^{t_2} f^2(t_1, \ldots, t_n)\, \mathrm{d}t_1 \ldots \mathrm{d}t_n < \infty.$$

Furthermore $\{\bar{Z}_n : n = 0, 1, \ldots\}$ is a sequence of mutually orthogonal closed subspaces in $L^2(\gamma)$, and Wiener's decomposition theorem asserts that

$$L^2(\gamma) = \bigoplus_{n=0}^{\infty} \bar{Z}_n.$$

By definition of $\mathscr{L}$ each Z_n is an eigenspace of $\mathscr{L}$ with eigenvalue $-(n/2)$, and extending this relationship to $\bar{Z}_n$ one arrives at the following description of $\mathscr{L}$. For each n let $\pi_n : L^2(\gamma) \to \bar{Z}_n$ be the orthogonal projection onto $\bar{Z}_n$. Then

$$\mathscr{D}(\mathscr{L}) = \left\{ \phi \in L^2(\gamma) : \sum_{n=1}^{\infty} n^2 \, \|\pi_n \phi\|_{L^2}^2 < \infty \right\}$$

and

$$\mathscr{L}\phi = \sum_{n=1}^{\infty} -\frac{n}{2} \pi_n \phi \quad \text{for} \quad \phi \in \mathscr{D}(\mathscr{L}).$$

Remark This linear operator is known in quantum physics as the number operator.

Theorem 2.4 *$\mathscr{L}$ is an SDO for γ.*

Proof Clearly $\mathscr{L}$ satisfies (i) and (ii) of Definition 2.1.

Define $\mathscr{D}$ as $\text{span}\{H_\alpha(\vec{w}) = H_{\alpha_1}(\vec{w}_1) \ldots H_{\alpha_n}(\vec{w}_n) : n \geq 0;\ \alpha_1, \ldots, \alpha_n$ are non-negative integers$\}$. By the preceding remarks $\mathscr{D} \subseteq \mathscr{D}(\mathscr{L})$ and $\mathscr{L}(\mathscr{D}) \subseteq \mathscr{D}$, and since $\vec{w}$ is a sequence of independent $N(0, 1)$ random variables it follows that $\mathscr{D}$ and hence $\mathscr{L}(\mathscr{D}) \subseteq L^4(\gamma)$. Furthermore, since for each $n \geq 0$ $\text{span}\{H_m : 0 \leq m \leq n\} = \text{span}\{x^m : 0 \leq m \leq n\}$, it is easy to check that if $\phi \in \mathscr{D}$ then $\phi^2 \in \mathscr{D}$ and hence $\phi^2 \in \mathscr{D}(\mathscr{L})$. Finally, suppose that

$$\phi = \sum_{n=0}^{\infty} \phi_n \in \mathscr{D}(\mathscr{L})$$

where each $\phi_n \in \bar{Z}_n$. For each i and $n = 0, 1, \ldots, i$ choose $\phi_n^i \in Z_n$ such that $\|\phi_n^i - \phi_n\| < 1/2^i$ (here $\|\cdot\|$ denotes the L^2 norm) and let

$$\phi^i = \sum_{n=0}^{i} \phi_n^i.$$

Then each $\phi_i \in \mathcal{D}$. Furthermore

$$\|\mathcal{L}\phi - \mathcal{L}\phi_i\| \leq \sum_{n=1}^{i} n \|\phi_n - \phi_n^i\| + \sum_{n=i+1}^{\infty} n \|\phi_n\|$$
$$\leq i(i+1)/2^{i+1} + \left\| \sum_{n=i+1}^{\infty} n\phi_n \right\| \to 0 \quad \text{as} \quad i \to \infty.$$

A similar calculation shows that $\|\phi - \phi^i\| \to 0$ as $i \to \infty$. This shows that graph$(\mathcal{L}/\mathcal{D})$ is dense in graph$(\mathcal{L})$, i.e. (iii) is satisfied.

We will need the following result in order to check (iv) and (v). Suppose that $n \geq 1$ and $F \in C^2_\uparrow(\mathbf{R}^n)$. Then $F(\vec{w}_1, \ldots, \vec{w}_n) \in \mathcal{D}(\mathcal{L})$ and

$$\mathcal{L}(F(\vec{w}_1, \ldots, \vec{w}_n))$$
$$= \frac{1}{2}\left[\sum_{i=1}^{n} \frac{\partial^2 F}{\partial x_1^2}(\vec{w}_1, \ldots, \vec{w}_n) - \sum_{i=1}^{n} \vec{w}_i \frac{\partial F}{\partial x_i}(\vec{w}_1, \ldots, \vec{w}_n)\right]. \quad (2.9)$$

To prove this, suppose first that $F(x_1, \ldots, x_n) = H_{\alpha_1}(x_1) \ldots H_{\alpha_n}(x_n)$. Then by Lemma 2.2

$$\frac{1}{2}\left[\sum_{i=1}^{n} \frac{\partial^2 F}{\partial x_1^2}(x) - \sum_{i=1}^{n} x_i \frac{\partial F}{\partial x_i}(x)\right] = -\frac{|\alpha|}{2} F(x).$$

Since $F(\vec{w}_1, \ldots, \vec{w}_n) \in Z_{|\alpha|}$, this proves (2.9) for F and hence for any function in span$\{H_{\alpha_1}(x_1) \ldots H_{\alpha_n}(x_n)\}$. The proof of (2.9) for general $F \in C^2_\uparrow(\mathbf{R}^n)$ now follows the same lines as that of Theorem (2.3) (iii), i.e. one uses denseness to approximate F by functions of the above form.

It now follows from (2.9) that if $\boldsymbol{\alpha} = (\alpha_1, \ldots, \alpha_m)$ and $\boldsymbol{\beta} = (\beta_1, \ldots, \beta_m)$, then

$$\langle H_{\boldsymbol{\alpha}}(\vec{w}), H_{\boldsymbol{\beta}}(\vec{w})\rangle = \sum_{i=1}^{\min(n,m)} \frac{\partial H_{\boldsymbol{\alpha}}}{\partial x_i}(\vec{w}) \frac{\partial H_{\boldsymbol{\beta}}}{\partial x_i}(\vec{w}).$$

This relation implies that $\langle , \rangle$ is a non-negative definite form on $\mathcal{D} \times \mathcal{D}$, so (iv) is satisfied.

Suppose F is as in part (v) of Definition (2.1) and each ϕ_i is of the form

$$H_{\boldsymbol{\alpha}^i}(\vec{w}) = H_{\alpha_1^i}(\vec{w}_1) \ldots H_{\alpha_{m_i}^i}(\vec{w}_{m_i}) \quad (2.10)$$

for $i = 1, \ldots, n$. Then

$$F(H_{\boldsymbol{\alpha}^1}(x_1, \ldots, x_{m_1}), \ldots, H_{\boldsymbol{\alpha}^n}(x_1, \ldots, x_{m_n})) \in C^2_\uparrow(\mathbf{R}^k)$$

where $k = \max(m_1, \ldots, m_n)$. Hence $F(\phi_1, \ldots, \phi_n) \in \mathscr{D}(\mathscr{L})$ and applying (2.9) yields

$$\begin{aligned}
&\mathscr{L}(F(\phi_1, \ldots, \phi_n)) \\
&= \frac{1}{2}\left\{\sum_{i,j=1}^{n} \langle H_{\alpha^i}(\vec{w}), H_{\alpha^j}(\vec{w})\rangle \frac{\partial^2 F}{\partial x_i\,\partial x_j}(H_{\alpha^1}(\vec{w}), \ldots, H_{\alpha^n}(\vec{w}))\right. \\
&\quad \left. + \sum_{i=1}^{n} \mathscr{L}(H_{\alpha^i}(\vec{w}))\frac{\partial F}{\partial x_i}(H_{\alpha^1}(\vec{w}), \ldots, H_{\alpha^n}(\vec{w}))\right\} \\
&= \frac{1}{2}\left\{\sum_{i,j=1}^{n} \langle \phi_i, \phi_j\rangle \frac{\partial^2 F}{\partial x_i\,\partial x_j}(\phi_1, \ldots, \phi_n)\right. \\
&\quad \left. + \sum_{i=1}^{n} \mathscr{L}\phi_i \frac{\partial F}{\partial x_i}(\phi_1, \ldots, \phi_n)\right\} \qquad (2.11)
\end{aligned}$$

as required. In general suppose that for $i = 1, \ldots, n$

$$\phi_i = \sum_{r=1}^{d_i} a_r^i \phi_r^i$$

where each ϕ_r^i is as in (2.10).

Write $F(\phi_1, \ldots, \phi_n)$ as $G(\phi_1^1, \ldots, \phi_{d_1}^1, \phi_1^2, \ldots, \phi_{d_2}^2, \ldots, \phi_1^n, \ldots, \phi_{d_n}^n)$ and apply (2.11) to G. Rewriting the result in terms of F shows that $F(\phi_1, \ldots, \phi_n)$ also satisfies (2.11). Hence $\mathscr{L}$ satisfies (v).

In order to check (vi) we introduce an infinite dimensional version of the Ornstein–Uhlenbeck process defined in the proof of Theorem 2.3(vi). Let $\{\theta_\tau\}_{\tau \ge 0}$ be Brownian motion with values in C_0, i.e. an a.s. continuous (with respect to the supremum norm on C_0) time homogeneous independent increment process such that $\theta_0 = 0$ a.s. We also assume that θ_1 has distribution γ and that $\{\theta_\tau\}_{\tau\ge 0}$ is independent of w where w now denotes a C_0-valued random variable with distribution γ. Define a path-valued process η by

$$\eta_\tau = w + \theta_\tau - \frac{1}{2}\int_0^\tau \eta_s \,\mathrm{d}s, \qquad \tau \ge 0. \tag{2.12}$$

As before this equation can be solved for η and the result will be

$$\eta_\tau = \mathrm{e}^{-\tau/2}\left(w + \int_0^\tau \mathrm{e}^{s/2}\,\mathrm{d}\theta_s\right). \tag{2.13}$$

Here we are using the theory of Itô integration with respect to Banach space-valued Brownian motion (see e.g. Metivier and Pellaumail)[23]; however, the stochastic integral in (2.13) could also be written as an ordinary Stieltjes integral by integration by parts. Suppose that $B = \{e_n\}_{n=1}^\infty$ is an orthonormal basis of the space $L_0^{2,1}$ such that $B \subseteq (C_0)^*$

(where $(C_0)^*$ is as in section 1.1). Applying each e_n to (2.13) gives

$$e_n(\eta_\tau) = e^{-\tau/2}\left\{e_n(w) + \int_0^\tau e^{s/2}\,de_n(\theta)_s\right\}.$$

Now $\{e_n(w)\}_{n=1}^\infty$ is a sequence of independent $N(0, 1)$ random variables and $\{e_n(\theta)\}_{n=1}^\infty$ a sequence of independent normalized Brownian motions in $\mathbf{R}$, each independent of the values of $e_n(w)$. It follows from the proof of Theorem 2.3(vi) that $e_1(\eta_\tau)$, $e_2(\eta_\tau), \ldots$ are independent $N(0, 1)$ random variables, and this implies that for each τ, η_τ has the same distribution as w, i.e. γ. The proof of (vi) now follows similar lines to that of Theorem 2.3.

Defining a semi-group on $C_b(C_0)$ (bounded functions on C_0) by $T_\tau(\phi)(w) = E[\phi(\eta_\tau)/w]$ it follows from the stationarity of η that the operators T_τ are contractions with respect to the $L^p(\gamma)$ norms; thus $\{T_\tau\}_{\tau\geq 0}$ extends to a contraction semi-group $\{T_\tau^p\}_{\tau\geq 0}$ on $L^p(\gamma)$, for each p. Let $\mathscr{L}_2$ be the generator of $\{T_\tau^2\}_{\tau\geq 0}$ and suppose that $\phi \in \mathscr{D}$. Then ϕ is of the form $F(\vec{w}_1, \ldots, \vec{w}_n)$ for some $n \geq 0$ and $F \in C_\uparrow^2(\mathbf{R}^n)$. Without loss of generality we can assume that the functions $\{f_n\}_{n=1}^\infty$ which define $\rightarrow$ are of bounded variation on $[0, 1]$. Then, since for each i one has

$$\vec{w}_i = \int_0^1 f_i\,dw = f_1 w_1 - \int_0^1 w\,df_i$$

it follows that the map $w \to \vec{w}_i \in (C_0)^*$. Hence $\phi \in C_\uparrow^2(C_0)$, and writing $\phi = F \circ P_n$ where $P_n(w) = (\vec{w}_1, \ldots, \vec{w}_n)$ and applying Itô's lemma to (2.12) shows that $\phi \in \mathscr{D}(\mathscr{L}_2)$ and

$$(\mathscr{L}_2\phi)(w) = \tfrac{1}{2}\{\mathrm{Tr}_{L_0^{2,1}} D^2\phi(w) - D\phi(w)w\}.$$

Evaluating the trace on the following basis of $L_0^{2,1}$

$$\left\{\int_0 f_1\,dt, \int_0 f_2\,dt, \ldots\right\}$$

gives

$$(\mathscr{L}_2\phi)(w) = \frac{1}{2}\left\{\sum_{i=1}^n \frac{\partial^2 F}{\partial x_i^2}(P_n w) - \sum_{i=1}^n \vec{w}_i \frac{\partial F}{\partial x_i}(P_n w)\right\}$$

and by (2.9) this coincides with $\mathscr{L}\phi$. Since $\mathscr{L}_2$ is closed and property (iii) holds, this implies that $\mathscr{L}_2$ extends $\mathscr{L}$. However, once again, $\{T_\tau^2\}_{\tau\geq 0}$ and hence $\mathscr{L}_2$ are self-adjoint, and since $\mathscr{L}$ is also self-adjoint, it follows that $\mathscr{L} = \mathscr{L}_2$. Hence $\mathscr{L}_1$ is a closed extension of $\mathscr{L}$ on $L^1(\gamma)$ and (vi) now follows from relation (2.8) as before. The proof of Theorem 2.4 is now complete. ■

Zakai makes the following interesting observation in his paper[29]. Suppose $\phi \in L^2(\gamma)$. Then ϕ can be extended to $\phi(\lambda\cdot)$ for $\lambda \in [0, 1]$ by defining

$$\phi(\lambda\cdot) = \sum_{n=0}^{\infty} \lambda^n \phi_n$$

where

$$\phi = \sum_{n=0}^{\infty} \phi_n$$

is the Wiener decomposition. With this extension it turns out that $\phi \in \mathscr{D}(\mathscr{L})$ if and only if

$$\tilde{D}\phi(w)w := \lim_{\lambda\uparrow 1} \frac{\phi(w) - \phi(\lambda w)}{1 - \lambda}$$

exists in L^2 in which case $(\mathscr{L}\phi)(w) = -1/2\tilde{D}\phi(w)w$. It should be noted, however, that in the case where ϕ has a continuous extension ϕ_c to C_0 (in the Banach norm), then $\phi(\lambda w)$ does not generally coincide with $\phi_c(\lambda w)$. As an example consider

$$\phi(w) = \int_0^1 w_s \, dw_s.$$

Then, since $\phi \in \bar{Z}_2$,

$$\phi(\lambda w) = \lambda^2 \int_0^1 w_s \, dw_s.$$

However, by Itô's lemma

$$\phi(w) = \frac{w_1^2}{2} - \frac{1}{2}$$

and since the latter formula defines ϕ_c, it is clear that $\phi_c(\lambda w) = \lambda^2 w_1^2/2 - (1/2) \neq \phi(\lambda w)$ if $0 < \lambda < 1$.

Suppose now that γ is the n-dimensional Wiener measure on $C_0([0, 1]; \mathbf{R}^n)$. In this case there exists the decomposition

$$L^2(\gamma) = \bigoplus_{m=0}^{\infty} \bar{Z}_m$$

where, for each m, $\bar{Z}_m$ is as defined in Theorem 1.16.

If one defines $\pi_m : L^2(\gamma) \to \bar{Z}_m$ to be orthogonal projection,

$$\mathscr{D}(\mathscr{L}) = \left\{ \phi \in L^2(\gamma) : \sum_{m=0}^{\infty} m^2 \|\pi_m \phi\|_{L^2}^2 < \infty \right\}$$

and

$$\mathscr{L}\phi \sum_{m=1}^{\infty} (-m/2)\pi_m\phi \qquad \text{for} \qquad \phi \in \mathscr{D}(\mathscr{L}),$$

then it can be shown that $\mathscr{L}$ is an SDO for γ. The arguments used in the proof of Theorem 2.4 generalize in an obvious way.

2.3 Regularity of measures induced by Wiener functionals

Our objective in this section is to show how the existence of an SDO $\mathscr{L}$ for γ leads to conditions under which the functional $\Phi = (\Phi_1, \ldots, \Phi_d) \in \mathscr{D}(\mathscr{L})^d$ induces an absolutely continuous distribution on $\mathbf{R}^d$ with a smooth density function. In fact, continuing to follow Stroock[26], we will prove the following result.

Theorem 2.5 *Suppose that $\Phi_1, \ldots, \Phi_d$ satisfy the following conditions:*
(i) *For any set $M = \{\psi_1, \ldots, \psi_m\} \subset \mathscr{D}(\mathscr{L})$, define $g(M) = \{\psi_i, \mathscr{L}\psi_i, \langle\psi_i, \psi_j\rangle : 1 \le i, j \le d\}$. Suppose that $M_0 \equiv \{\phi_1, \ldots, \phi_d\} \subset \mathscr{D}(\mathscr{L})$, $M_1 \equiv g(M_0) \subset \mathscr{D}(\mathscr{L})$, $M_2 \equiv g(M_1) \subset \mathscr{D}(\mathscr{L}), \ldots,$* etc.
(ii) *For each $r \ge 0$,*

$$M_r \subset \bigcap_{p=1}^{\infty} L^p(\gamma).$$

(iii) *Define the matrix σ by $\sigma_{ij} = \langle\Phi_i, \Phi_j\rangle$, $1 \le i, j \le d$ and let $\Delta = \det \sigma$. Suppose that*

$$\frac{1}{\Delta} \in \bigcap_{p=1}^{\infty} L^p(\gamma).$$

Then the distribution of $(\Phi_1, \ldots, \Phi_d)$ is absolutely continuous with respect to the Lebesgue measure on $\mathbf{R}^d$ and has a C^∞ density.

The proof of Theorem 2.5 will require the following lemma. The proof of the lemma (which we omit) is quite straightforward and utilizes the properties of $\mathscr{L}$ given in Definition 2.1 (cf. Stroock[26], pp. 400, 401).

Lemma 2.6 *The form $\langle , \rangle$ admits a unique extension to $\mathscr{D}(\mathscr{L}) \times \mathscr{D}(\mathscr{L})$ so that $(\phi, \psi) \in \mathscr{D}(\mathscr{L})^2 \to \langle\phi, \psi\rangle \in L^1(\gamma)$ is a non-negative bilinear symmetric map which is continuous with respect to the graph($\mathscr{L}$)-norm. Suppose that $a_1, \ldots, a_n, b \in \mathscr{D}(\mathscr{L})$ and $F \in C_b^2(\mathbf{R}^n)$. Then $F \circ a \in \mathscr{D}(\mathscr{L}_1)$,* *where* $a = (a_1, \ldots, a_n)$,

$$\mathscr{L}_1(F \circ a) = \frac{1}{2} \sum_{i,j=1}^{n} \langle a_i, a_j \rangle \frac{\partial^2 F}{\partial x_i \partial x_j} \circ a + \sum_{i=1}^{n} \mathscr{L} a_i \frac{\partial F}{\partial x_i} \circ a \tag{2.14}$$

and if $\mathscr{L}_1(F \circ a) \in L^2(\gamma)$, *then*

$$\langle F \circ a, b \rangle = \sum_{i=1}^{n} \frac{\partial F}{\partial x_i} \circ a \langle a_i, b \rangle \tag{2.15}$$

Furthermore, if $a, b \in \mathscr{D}(\mathscr{L})$, *then* $ab \in \mathscr{D}(\mathscr{L}_1)$ *and*

$$\mathscr{L}_1(ab) = a\mathscr{L}b + b\mathscr{L}a + \langle a, b \rangle \tag{2.16}$$

Lemma 2.7 *Suppose that* $a, b, c \in \mathscr{D}(\mathscr{L})$ *have the properties:* $a > 0$ *a.s., and* $1/a$, a, b, c, $\mathscr{L}a$, $\mathscr{L}b$, $\mathscr{L}c$, $\langle a, a \rangle$, $\langle b, b \rangle$, $\langle c, c \rangle$ *lie in* $\bigcap_{p=1}^{\infty} L^p$. *Then*

(i) $1/a \in \mathscr{D}(\mathscr{L})$ *and* $\mathscr{L}(1/a) = \langle a, a \rangle / a^3 - \mathscr{L}a/a$

(ii) $\langle 1/a, b \rangle = -1/a^2 \langle a, b \rangle$

(iii) $ab \in \mathscr{D}(\mathscr{L})$, $\mathscr{L}(ab) = a\mathscr{L}b + b\mathscr{L}a + \langle a, b \rangle$ *and* $\langle ab, c \rangle = a\langle b, c \rangle + b\langle a, c \rangle$.

Proof (i) For each $\varepsilon > 0$, let $F_\varepsilon(x) = (x^2 + \varepsilon^2)^{-1/2}$. Then $F_\varepsilon \in C_b^2(\mathbf{R})$. Hence $F_\varepsilon(a) \in \mathscr{D}(\mathscr{L}_1)$ and by (2.14)

$$\begin{aligned}\mathscr{L}_1(F_\varepsilon(a)) = \tfrac{1}{2}\langle a, a \rangle (3a^2(a^2 + \varepsilon^2)^{-5/2} - (a^2 + \varepsilon^2)^{-3/2}) \\ -(\mathscr{L}a)(a^2 + \varepsilon^2)^{-3/2} \cdot a\end{aligned} \tag{2.17}$$

However, since the right-hand side is in L^2, $\mathscr{L}_1$ can be replaced by $\mathscr{L}$. Now let $\varepsilon \downarrow 0$. Then $F_\varepsilon(a) \to 1/a$ and the right-hand side of (2.17) tends to $\langle a, a \rangle / a^3 - (\mathscr{L}a)/a$ in L^2 and since $\mathscr{L}$ is closed (i) follows. Part (ii) can be proved in the same way, and part (iii) follows from (2.15) and (2.16). ■

Corollary 2.8 From the inequality $|\langle a, b \rangle| \leq \langle a, a \rangle^{1/2} \langle b, b \rangle^{1/2}$ it follows that in (iii) of Lemma 2.7, $\mathscr{L}(ab)$ and $\langle ab, c \rangle \in \bigcap_{p=1}^{\infty} L^p$.

Remark Suppose that $a_1, \ldots, a_r$; $b_1, \ldots, b_s$; $c_1, \ldots, c_t$ satisfy the conditions on a, b and c respectively in the above lemma, and R and S are polynomials in $a_1^{-1}, \ldots, a_r^{-1}$, $b_1, \ldots, b_s$, $c_1, \ldots, c_t$. Then iterating Lemma 2.7 shows that R and S are in $\mathscr{D}(\mathscr{L})$ and $\mathscr{L}R$ and $\langle R, S \rangle$ are in L^p for each $p \geq 1$.

We are now in a position to prove Theorem 2.5. Suppose that ϕ, b and $y_1, \ldots, y_b$ are as in Lemma 1.14, and R is as in the above remark, where $a_1, \ldots, a_r$, $b_1, \ldots, b_s$, $c_1, \ldots, c_t$ are functionals of the type that appear in the sets M_r in Theorem 2.5(i). Let $(A_{ij})_{i,j=1}^d$ be the cofactor matrix of σ.

Then we have the following relation

$$E[D^b\phi(\Phi)(y_1, \ldots, y_b)\cdot R] = E\left[\frac{R}{\Delta}\sum_{i,j=1}^{d} y_b^i A_{ij}\langle D^{b-1}\phi(\Phi)(y_1, \ldots, y_{b-1}), \Phi_j\rangle\right] \tag{2.18}$$

where y_b^i, $i = 1, \ldots, d$ are the components of y_b. This is easily verified by applying (2.15) with $F \equiv D^{b-1}\phi(\cdot)(y_1, \ldots, y_{b-1})$, $a \equiv \Phi$ and $b \equiv \phi_j$. Note that Δ and each A_{ij} are polynomials in the components of σ so by Corollary 2.8 it follows that $A_{ij}R/\Delta \in \mathcal{D}(\mathcal{L})$ and using Lemma 2.7(iii) and then the symmetry of $\mathcal{L}$ on $L^2(\gamma)$, we may write the above in the form

$$\begin{aligned} E\Bigg[\sum_{i,j=1}^{d} & y_b^i\{\mathcal{L}(D^{b-1}\phi(\Phi)(y_1, \ldots, y_{b-1})\Phi_j) \\ & - D^{b-1}\phi(\Phi)(y_1, \ldots, y_{b-1})\mathcal{L}\Phi_j \\ & - \Phi_j\mathcal{L}(D^{b-1}\phi(\Phi)(y_1, \ldots, y_{b-1}))\}RA_{ij}/\Delta\Bigg] \\ = E\Bigg[\sum_{i,j=1}^{d} & y_b^i D^{b-1}\phi(\Phi)(y_1, \ldots, y_{b-1}) \\ & \times\{\Phi_j\mathcal{L}(RA_{ij}/\Delta) - (\mathcal{L}\Phi_j)RA_{ij}/\Delta - \mathcal{L}(\Phi_j RA_{ij}/\Delta)\}\Bigg]. \end{aligned}$$

However, by Lemma 2.7 the expression in the brackets { } can be written as

$$\begin{aligned} \phi_j\{&\mathcal{L}(RA_{ij})/\Delta + RA_{ij}(\langle\Delta, \Delta\rangle/\Delta^3 - \mathcal{L}\Delta/\Delta) \\ &- 1/\Delta^2\langle RA_{ij}, \Delta\rangle\} + \text{two similar terms} \end{aligned}$$

and this is of the same form as R. Hence we may iterate this step, and successive iteration from (2.18) with $R = 1$ will eventually lead to the relation

$$E[D^b\phi(\Phi)(y_1, \ldots, y_b)] = E[\phi\circ\Phi\cdot S]$$

where $S \in L^1(\gamma)$. Hence the conclusion of the theorem follows from Lemma 1.14.

2.4 Application of the Malliavin calculus to stochastic differential equations

In this section (Theorem 2.10) it will be shown that the solutions of stochastic differential equations with suitably regular coefficients satisfy hypotheses (i) and (ii) of Theorem 2.5, i.e. that the Malliavin operations can be applied arbitrarily many times and the result will always be a functional in $L^p(\gamma)$ for every p. A stochastic differential equation will be obtained for the covariance matrix σ. However, the verification of hypothesis (iii) in Theorem 2.5 under Hörmander's condition involves some delicate analysis, and this will be delayed until chapter 6.

Suppose that w is normalized Brownian motion in $\mathbf{R}^n$ and that $\mathcal{L}$ is the SDO for the n-dimensional Wiener measure γ which we defined in section 2.2. For each $t \in [0, 1]$, let $\mathcal{F}_t = \sigma(w_s : s \le t)$ and for each $p \ge 1$, define

$$\mathcal{D}_p = \{\Phi \in \mathcal{D}(\mathcal{L})/ \\ \|\Phi\|_{(p)} \equiv (\|\Phi\|^p_{L^p} + \|\mathcal{L}\Phi\|^p_{L^p} + \|\langle \Phi, \Phi\rangle^{1/2}\|^p_{L^p})^{1/p} < \infty\}$$

We will require the following.

Theorem 2.9 *Suppose that* $\alpha_1 = (\alpha_1^1, \ldots, \alpha_1^n)$, $\alpha_2 = (\alpha_2^1, \ldots, \alpha_2^n)$: $[0, 1] \to \mathcal{D}^n$ *and* $\beta_1, \beta_2 : [0, 1] \to \mathcal{D}$ *are adapted and also*

$$E\left[\int_0^1 (\|\alpha_i^k(t)\|^p_{(p)} + \|\beta_i(t)\|^p_{(p)})\, dt\right] < \infty \tag{2.19}$$

for $k = 1, \ldots, n$ *and* $i = 1, 2$.

Let

$$x_i(t) = \int_0^t \alpha_i(s) \cdot dw_s + \int_0^t \beta_i(s)\, ds; \qquad t \in [0, 1]$$

for $i = 1, 2$.

If (2.19) *holds with* $p = 2$, *then* $x_i(t) \in \mathcal{D}(\mathcal{L})$, $i = 1, 2$. *Furthermore, if we define* $(\mathcal{L}\alpha_i)(s) = (\mathcal{L}\alpha_i^1(s), \ldots, \mathcal{L}\alpha_i^n(s))$ *and* $\langle \alpha_i(s), \cdot\rangle = (\langle \alpha_i^1(s), \cdot\rangle, \ldots, \langle \alpha_i^n(s), \cdot\rangle)$ *then*

$$\mathcal{L}x_i(t) = \int_0^t (\mathcal{L}\alpha_i - \tfrac{1}{2}\alpha_i)(s) \cdot dw_s + \int_0^t \mathcal{L}\beta_i(s)\, ds, \qquad t \in [0, 1]. \tag{2.20}$$

Finally, if (2.19) *holds with* $p = 4$, *then*

$$\begin{aligned}\langle x_i(t), x_j(t)\rangle = \int_0^t (\langle \alpha_i(s), x_j(s)\rangle + \langle \alpha_j(s), x_i(s)\rangle) \cdot dw_s \\ + \int_0^t \Big\{\langle \beta_i(s), x_j(s)\rangle + \langle \beta_j(s), x_i(s)\rangle \\ + \sum_{k=1}^n \langle \alpha_i^k(s), \alpha_j^k(s)\rangle \\ + \alpha_i(s) \cdot \alpha_j(s)\Big\}\, ds, \qquad t \in [0, 1].\end{aligned} \tag{2.21}$$

Proof Suppose first that ϕ and ψ are $\mathcal{F}_t$-measurable, in $\mathcal{D}(\mathcal{L})$ and $\Phi \in \bar{Z}_m$ is of the form

$$\int_0^1 \int_0^{t_m} \cdots \int_0^{t_2} f(t_1, \ldots, t_m)\, dw_{j_1}(t_1) \ldots dw_{j_m}(t_m);$$

$$1 \le j_1, \ldots, j_m \le n.$$

The $\mathcal{F}_t$-measurability of ϕ implies that $f(t_1, \ldots, t_m) = 0$ if $t_m > t$. Hence if $1 \le k \le n$ and $h > 0$, then defining

$$g(t_1, \ldots, t_{m+1}) = f(t_1, \ldots, t_m)\chi_{[t,t+h]}(t_{m+1}),$$

we can write

$$\phi \cdot (w_k(t+h) - w_k(t)) = \int_0^1 \int_0^{t_{m+1}} \cdots \int_0^{t_2} g(t_1, \ldots, t_{m+1})\, dw_{j_1}(t_1) \ldots dw_{j_m}(t_m)\, dw_k(t_{m+1})$$

and it follows that $\phi \cdot (w_k(t+h) - w_k(t)) \in \bar{Z}_{m+1}$. This implies

$$\begin{aligned}\mathcal{L}(\phi \cdot (w_k(t+h) - w_k(t))) &= -\left(\frac{m+1}{2}\right)\phi \cdot (w_k(t+h) - w_k(t)) \\ &= ((\mathcal{L} - \tfrac{1}{2})\phi)(w_k(t+h) - w_k(t)) \qquad (2.22)\end{aligned}$$

and since $\mathcal{L}$ is closed on $L^2(\gamma)$ this relation will hold for every $\mathcal{F}_t$ measurable $\phi \in \mathcal{D}(\mathcal{L})$. Equation (2.22) implies that if S is any adapted simple function with components in $\mathcal{D}(\mathcal{L})$, then

$$\int_0^t S\, dw_k \in \mathcal{D}(\mathcal{L}) \quad \text{and} \quad \mathcal{L}\left(\int_0^t S\, dw_k\right) = \int_0^t (\mathcal{L}S - S/2)\, dw_k \qquad (2.23)$$

The hypotheses on α_i imply that for each $1 \le k \le n$ there exists a sequence of simple functions $\{S_m\}$ of the above form such that

$$\int_0^t S_m\, dw_k \xrightarrow{L^2} \int_0^t \alpha_i^k\, dw_k \quad \text{and} \quad \int_0^t (\mathcal{L}S_m)\, dw_k \xrightarrow{L^2} \int_0^t \mathcal{L}\alpha_i^k\, dw_k$$

and again using the closed property of $\mathcal{L}$ we conclude that (2.23) holds for each k with α_i^k in place of S. A similar argument shows that

$$\int_0^t \beta_i(s)\, ds \in \mathcal{D}(\mathcal{L}) \quad \text{and} \quad \mathcal{L}\left(\int_0^t \beta_i(s)\, ds\right) = \int_0^t \mathcal{L}\beta_i(s)\, ds$$

and (2.20) follows.

To prove the second assertion one simply computes $x_i(t)x_j(t)$ from Itô's lemma, observes that the integrals obtained satisfy condition (2.19) with

$p=2$, and then evaluates $\mathscr{L}(x_i(t)x_j(t))$ by (2.20). Then computing $x_i(t)\mathscr{L}x_j(t)$ and $x_j(t)\mathscr{L}x_i(t)$ also by Itô's lemma and evaluating $\langle x_i(t), x_j(t)\rangle$ by the formula $\mathscr{L}(x_i(t)x_j(t)) - x_i(t)\mathscr{L}x_j(t) - x_j(t)\mathscr{L}x_i(t)$ will lead to (2.21). ■

Theorem 2.10 *Suppose that $x \in \mathbf{R}^d$, $A:\mathbf{R}^d \to \mathbf{M}^{n,d}$ and $B:\mathbf{R}^d \to \mathbf{R}^d$ are bounded, C^∞, and have bounded derivatives of all orders. Let ξ be the solution of the stochastic equation*

$$\xi_t = x + \int_0^t A(\xi_s)\,dw_s + \int_0^t B(\xi_s)\,ds; \qquad t \in [0, 1] \tag{2.24}$$

Then for each $t \in [0, 1]$, $\xi_t = (\xi_1(t), \ldots, \xi_d(t))$ satisfies conditions (i) *and* (ii) *in Theorem 2.5. Furthermore the covariance matrix $\sigma(t) = (\langle \xi_i(t), \xi_j(t)\rangle)_{i,j=1}^d$ satisfies the equation*

$$\begin{aligned}\sigma(t) = \sum_{k=1}^n \int_0^t [S_k(\xi_s)\sigma(s) + \sigma(s)S_k(\xi_s)^*]\,dw_k(s) \\ + \int_0^t \{C(\xi_s)\sigma(s) + \sigma(s)C(\xi_s)^* \\ + \sum_{k=1}^n S_k(\xi_s)\sigma(s)S_k(\xi_s)^* + A(\xi_s)A(\xi_s)^*\}\,ds\end{aligned} \tag{2.25}$$

where

$$S_k(x) = \left(\frac{\partial A_{ik}}{\partial x_j}(x)\right)_{i,j=1}^d,$$

and

$$C(x) = \left(\frac{\partial B_i}{\partial x_j}(x)\right)_{i,j=1}^d.$$

Proof The theorem is proved by using the Picard iteration scheme. Let θ denote the space of adapted processes from $[0, 1]$ into $\mathbf{R}^d$ with finite $\|\cdot\|_2^*$ norm, where

$$\|y\|_p^* = \sqrt[p]{\left(\sup_{t\in[0,1]} \|y(t)\|_{\mathbf{R}^d}^p\right)}.$$

Define a map $T:\theta \to \theta$ by

$$T(y)_t = x + \int_0^t A(y_s)\,dw_s + \int_0^t B(y_s)\,ds; \qquad t \in [0, 1]$$

for $y \in \theta$. Then define a sequence $\{\xi^{(m)}\}_{m\geq 1}$ inductively by $\xi^{(1)} \equiv x$ and

$\xi^{(m+1)} = T(\xi^{(m)})$, $m \geq 1$. It can be shown that $\xi^{(m)} \to \xi$ in $\|\cdot\|_p^*$ for all p (in fact this is the method used to show that equation (2.24) has a solution; see, for example, Bichteler).[4]

Applying Theorem 2.9 inductively to the equations

$$\xi_t^{(m+1)} = x + \int_0^t A(\xi_s^{(m)})\,dw_s + \int_0^t B(\xi_s^{(m)})\,ds$$

shows that $\xi_i^{(m)} \in \mathscr{D}(\mathscr{L})$ for each m and $1 \leq i \leq d$ and if we define $\sigma_t^{(m)} = (\langle \xi_i^{(m)}(t), \xi_j^{(m)}(t) \rangle)_{i,j=1}^d$ then (2.21) together with (2.15) yields the following relations:

$$\begin{aligned}
\sigma_t^{(m+1)} = & \int_0^t [S_k(\xi_s^{(m)})\sigma_s^{(m)} + \sigma_s^{(m)} S_k(\xi_s^{(m)})^*]\,dw_s \\
& + \int_0^t \Big\{ C(\xi_s^{(m)})\sigma_s^{(m)} + \sigma_s^{(m)} C(\xi_s^{(m)})^* \\
& + \sum_{k=1}^n S_k(\xi_s^{(m)})\sigma(s) S_k(\xi_s^{(m)})^* + A(\xi_s^{(m)})A(\xi_s^{(m)})^* \Big\}\,ds
\end{aligned}$$

$$\begin{aligned}
\mathscr{L}\xi_i^{(m+1)}(t) = x + & \sum_{k=1}^n \int_0^t \Big\{ \frac{1}{2} \sum_{p,q=1}^d \sigma_{p,q}^{(m)}(s) \frac{\partial^2 A_{ik}}{\partial x_p\,\partial x_q}(\xi_s^{(m)}) \\
& + \sum_{p=1}^d \mathscr{L}\xi_s^{(m)} \frac{\partial A_{ik}}{\partial x_p}(\xi_s^{(m)}) - \tfrac{1}{2} A_{ik}(\xi_s^{(m)}) \Big\}\,dw_k(s) \\
& + \int_0^t \Big\{ \frac{1}{2} \sum_{p,q=1}^d \sigma_{pq}^{(m)}(s) \frac{\partial^2 B_i}{\partial x_p\,\partial x_q}(\xi_s^{(m)}) \\
& + \sum_{p=1}^d \mathscr{L}\xi_s^{(m)} \frac{\partial B_i}{\partial x_p}(\xi_s^{(m)}) \Big\}\,ds
\end{aligned}$$

Using the first relation, one can show (by the same method as that used to show that the original approximations $\xi^{(m)} \to \xi$) that $\sigma^{(m)}$ converges to a limit, say τ, in the norm $\|\cdot\|_p^*$ for every p and that τ satisfies equation (2.25). Using this fact, together with the second relation, it can be shown that $\mathscr{L}\xi^{(m)}$ converges to a limit η, again in $\|\ \|_p^*$ for all p. In particular $\mathscr{L}\xi_t^{(m)} \to \eta_t$ for every t, in $L^2(\gamma)$. Since $\xi_t^{(m)} \to \xi_t$ in $L^2(\gamma)$ and $\mathscr{L}$ is closed, it follows that $\xi_t \in \mathscr{D}(\mathscr{L})$ for every t and $\mathscr{L}(\xi_t^{(m)}) \to \mathscr{L}(\xi_t)$. (Note that $\mathscr{L}(\xi_t)$ satisfies the equation obtained from the second relation by deleting m.) The continuity of $\langle , \rangle$ in the graph($\mathscr{L}$)-norm now implies that the matrix τ above is in fact σ. We have therefore shown that σ satisfies (2.25). Furthermore, since ξ, $\mathscr{L}\xi$ and σ were obtained as limits of sequences in $\|\ \|_p^*$ for every p, it follows that ξ_i, $\mathscr{L}\xi_i$, $\langle \xi_i, \xi_j \rangle \in \bigcap_{p=1}^{\infty} L^p$ for $1 \leq i, j \leq d$.

Applying the same argument to the derived stochastic differential equations for σ and ξ, one concludes that the functionals above are also in $\mathcal{D}(\mathcal{L})$, and applying $\mathcal{L}$ and $\langle\ \rangle$ again yields elements of $\bigcap_{p=1}^{\infty} L^p$ and stochastic differential equations. This procedure may be repeated arbitrarily often. Thus the theorem is proved. ■

3

The variational approach

3.1 Perturbation via the Girsanov theorem

The work described in this chapter is actually a simplified version of Bismut's variational form of the Malliavin calculus[6] due to Bichteler and Fonken[5].

The underlying idea is as follows. Given Brownian motion† w defined with respect to a probability measure P, a process w^v is constructed for every $v \in \mathbf{R}$ by adding to w a perturbing term which depends smoothly on v. More specifically w^v is defined by

$$w_t^v = w_t - \int_0^t H_s v \, \mathrm{d}s, \qquad t \in [0, 1]$$

where H is a pre-selected bounded adapted process. It follows from the Girsanov theorem that for each v the distribution of w^v with respect to the measure P^v is the same as that of w with respect to P, where $\mathrm{d}P^v = G(v)\,\mathrm{d}P$ and $G(v)$ is the Girsanov density, given by

$$G(v) = \exp\left\{ v \int_0^1 H_s \, \mathrm{d}w_s - \frac{v^2}{2} \int_0^1 H_s^2 \, \mathrm{d}s \right\}.$$

Hence if F is any bounded measurable function of $\{w_s : 0 \le s \le 1\}$, then for every v the distributions of $F(w^v)$ with respect to P^v and $F(w)$ with respect to P will be the same. In particular we will have

$$\begin{aligned} E[F(w)] &= E_v[F(w^v)]\ddagger \\ &= E[F(w^v)G(v)] \end{aligned} \qquad (3.1)$$

Thus the quantity in (3.1) will be independent of v.

† For the sake of simplicity all processes in this section will be assumed to take values in $\mathbf{R}$. Sections 3.2 and 3.3 deal with higher dimensional processes.

‡ Here E and E_v denote expectations with respect to P and P^v respectively.

Suppose ξ is a process defined by a stochastic differential equation driven by w and let ξ^v denote the process obtained by replacing w by w^v throughout. Thus if ξ is the solution of the following equation

$$\xi_t = x + \int_0^t A(\xi_s)\, dw_s + \int_0^t B(\xi_s)\, ds, \qquad t \in [0, 1]$$

then ξ^v will satisfy

$$\begin{aligned}\xi_t^v &= x + \int_0^t A(\xi_s^v)\, dw_s^v + \int_0^t B(\xi_s^v)\, ds \\ &= x + \int_0^t A(\xi_s^v)\, dw_s + \int_0^t \{-A(\xi_s^v)H_s v + B(\xi_s^v)\}\, ds, \qquad t \in [0, 1].\end{aligned} \tag{3.2}$$

The results of section 3.2 establish that under smoothness conditions on A and B the map $v \to \xi^v$ is differentiable in an appropriate sense. Furthermore the process η given by

$$\eta = \frac{\partial}{\partial v} \xi^v \Big|_{v=0}$$

satisfies another stochastic differential equation driven by w, and if we denote by L the operator defined by $L(\xi) = \eta$, then L may be applied to ξ arbitrarily many times.

In order to derive regularity results about the distribution of ξ_t, $t > 0$, we need the condition

$$\eta_t \neq 0 \quad \text{a.s.} \tag{3.3}$$

Assuming this, let ϕ be a test function on $\mathbf{R}$ as in Lemma 1.14 and suppose that ψ is any measurable function of w such that $\psi(w^v)$ is differentiable with respect to v. Then choosing $F(w)$ in (3.1) to be $\psi(w)\phi(\xi_t)/\eta_t$ one concludes that

$$E[\{\psi(w^v)\phi(\xi_t^v)/\eta_t^v\}G(v)] \tag{3.4}$$

is independent of v. Differentiating with respect to v, evaluating at $v = 0$ and then equating to zero gives

$$E[\phi'(\xi_t)\psi(w)] = -E\left[\phi(\xi_t)\frac{\partial}{\partial v}\left\{\frac{G(v)\psi(w^v)}{\eta_t^v}\right\}\Big|_{v=0}\right]$$

The inequality (1.7) required to establish the absolute continuity of the distribution of ξ_t can now be obtained from this. Taking $\psi \equiv 1$ and then showing that

$$\frac{\partial}{\partial v}\left\{\frac{G(v)}{\eta_t^v}\right\}\bigg|_{v=0}$$

(which we will denote by $\tilde{\psi}$) has finite L^1-norm yields

$$|E[\phi'(\xi_t)]| \leq \|\phi\|_\infty \cdot \|\tilde{\psi}\|_{L^1}.$$

This scheme is amenable to iteration as follows. Replacing ϕ by ϕ' gives

$$E[\phi''(\xi_t)] = -E[\phi'(\xi_t)\tilde{\psi}] \tag{3.5}$$

and then repeating the argument with the right-hand side of (3.5), using $\tilde{\psi}$ in place of ψ, gives the relation

$$E[\phi''(\xi_t)] = E\left[\phi(\xi_t)\frac{\partial}{\partial v}\left\{\frac{G(v)\tilde{\psi}(w^v)}{\eta_t^v}\right\}\right]\bigg|_{v=0}.$$

An easy computation shows that the expression

$$\frac{\partial}{\partial v}\left\{\frac{G(v)\tilde{\psi}(w^v)}{\eta_t^v}\right\}\bigg|_{v=0}$$

is a polynomial in the terms $LG'(0)$, $(L\xi)_t$, $(L^2\xi)_t$, $(L^3\xi)_t$ and $1/\eta_t$.

In general we will have relations of the form

$$E[\phi^{(n)}(\xi_t)] = E[\phi(\xi_t)R_n(w)]$$

where R_n is a polynomial in $G'(0)$, $LG'(0), \ldots, L^{n-1}G'(0)$, $(L\xi)_t, \ldots, (L^{n+1}\xi)_t$ and $1/\eta_t$.

The higher order estimates (1.8) needed to prove smoothness of the density of ξ_t are obtained by showing that each of the terms above is in L^p for every p, which then implies that each of the polynomials R_n is in L^1.

The particular choice of the process H is determined by the requirement that η_t be non-zero. It turns out that η satisfies the equation

$$\eta_s = \int_0^s DA(\xi_u)\eta_u\, dw_u + \int_0^s \{-A(\xi_u)H_u + DB(\xi_u)\eta_u\}\, du, \qquad s \in [0, 1]$$

Furthermore η may be expressed in terms of the solution Y of the equation†

$$Y_s = 1 + \int_0^s DA(\xi_u)Y_u\, dw_u + \int_0^t DB(\xi_u)Y_u\, du$$

† The solution of a linear stochastic differential equation of this type is a.s. non-zero at every time s and Y_s and $1/Y_s$ are in L^p for all p.

as

$$\eta_s = -Y_s \int_0^s Y_u^{-1} A(\xi_u) H_u \, du \tag{3.6}$$

This formula suggests $A(\xi_u)Y_u$ as a natural choice for H_u since this will make the integrand in (3.6) non-negative. Hence at $s = t$ the integral will be strictly positive if $A(\xi_u)$ is non-zero for some $u \in [0, t]$, which in turn will imply that $\eta_t \neq 0$. Furthermore the condition $A(\xi_u) \neq 0$ for some $u \in [0, 1]$ a.s. is clearly necessary for ξ_t to have a density with respect to the Lebesgue measure.† (In the multi-dimensional case one chooses H_u to be $A(\xi_u)^*(Y_u^{-1})^*$ in order to give the integrand in (3.6) a non-negative definite form.) With this choice of H it can be shown that the required non-vanishing property of η_t (non-degeneracy in higher dimensions) together with the existence of the above L^p-norms are implied by Hörmander's conditions on A and B. This is done in chapter 6.

3.2 Smooth dependence of stochastic differential equations upon parameters

For any Euclidean space $\mathbf{R}^n$, let $B_p = B_p(\mathbf{R}^n)$ denote the Banach space of adapted processes with values in $\mathbf{R}^n$, with finite $\| \; \|_p$-norm, where $\|X\|_p$ is defined by

$$\left\| \sup_{s \in [0,1]} \|X_s\|_{\mathbf{R}^n} \right\|_{L^p}.$$

Throughout this section N will denote a (fixed) bounded neighbourhood of 0 in $\mathbf{R}^d$. We will say that a map X from N into B_p is C_p^1 if X is differentiable into B_p, and C_∞^1 if X is C_p^1 for every $p \in \mathbf{N}$.

Theorem 3.1 *Suppose that* $x: N \to \mathbf{R}^d$, $a: N \times \mathbf{R}^d \to \mathbf{M}^{n,d}$, $b: N \times \mathbf{R}^d \to \mathbf{R}^d$ *are bounded, twice differentiable maps with bounded first and second derivatives.*

For each $v \in N$ *let* X^v *denote the solution of the following stochastic differential equation driven by n-dimensional Brownian motion w*

$$X_t^v = x(v) + \int_0^t a(v, X_s^v) \, dw_s + \int_0^t b(v, X_s^v) \, ds, \qquad t \in [0, 1]. \tag{3.7}$$

† Since $A(\xi_u) \equiv 0$ with probability $\varepsilon > 0$ would imply $P(\xi_t = r_t) \geq \varepsilon$, where r is the solution of the non-random integral equation,

$$r_s = x + \int_0^s B(r_u) \, du, \qquad s \in [0, t].$$

Thus ξ_t would have a point mass at r_t.

Then the map $v \to X^v$ is C^1_∞. Furthermore for any $h \in \mathbf{R}^d$, $(D_v X^v)h$ is the solution η of the stochastic differential equation obtained by formal differentiation with respect to v in (3.7); i.e.

$$\eta_t = Dx(v)h + \int_0^t Da(v, X^v_s)(h, \eta_s)\,dw_s + \int_0^t Db(v, X^v_s)(h, \eta_s)\,ds,\dagger \qquad t \in [0, 1] \tag{3.8}$$

Proof We must show that the map $v \to X^v$ is C^1_p for every p. Without loss of generality it is sufficient to show this for every even p. Since the Cauchy–Schwartz inequality implies that $\| \ \|_p \le \| \ \|_{2p}$, the result will then follow for all p. In the following argument $C_1, C_2, \ldots, C_{16}$ will denote constants.

First note that X^v is in each B_p as follows. Applying the triangle inequality, then Theorem 1.9(iv) and Holder's inequality to (3.1) gives

$$\begin{aligned} \|X^v\|_p &\le x(v) + \left\| \int_0^\cdot a(v, X^v_s)\,dw_s \right\|_p + \left\| \int_0^\cdot b(v, X^v_s)\,ds \right\|_p \\ &\le x(v) + C_1 \left\{ \int_0^1 E\,\|a(v, X^v_s)\|_2^p\,ds \right\}^{1/p} + C_2 \left\{ \int_0^1 E\,\|b(v, X^v_s)\|^p\,ds \right\}^{1/p}. \end{aligned}$$

These last integrals are finite since a and b are bounded. Hence $X^v \in B_p$.

The next step is to show that $v \to X^v$ is Lipschitz into B_p, i.e. there exists a constant K such that

$$\|X^{v+h} - X^v\|_p \le K\,\|h\|. \tag{3.9}$$

We have $X^{v+h}_s - X^v_s$

$$= x(v+h) - x(v) + \int_0^t \{a(v+h, X^{v+h}_s) - a(v, X^v_s)\}\,dw_s + \int_0^t \{b(v+h, X^{v+h}_s) - b(v, X^v_s)\}\,ds.$$

† Here Da and Db denote the Frechet derivatives of a and b on $N \times \mathbf{R}^d$.

For $X \in B_p$, let $\|X\|_{p,t}$ denote

$$\left\| \sup_{s \in [0,t]} \|X(s)\|_{\mathbf{R}^d} \right\|_{L^p}.$$

The conditions on x, a and b imply that these are Lipschitz maps on $\mathbf{R}^d$ and $N \times \mathbf{R}^d$ respectively. Using this, together with Theorem 1.9(iv) and Holder's inequality, we obtain

$$\begin{aligned}\|X^{v+h} - X^v\|_{p,t}^p &\leq C_3 \|h\|^p + C_4 \left\| \int_0^{\cdot} \{a(v+h, X_s^{v+h}) - a(v, X_s^v)\} \, dw_s \right\|_{p,t}^p \\ &\quad + C_5 \left\| \int_0^{\cdot} \{b(v+h, X_s^{v+h}) - b(v, X_s^v)\} \, dw_s \right\|_{p,t}^p \\ &\leq C_3 \|h\|^p + C_6 \int_0^t E \, \|a(v+h, X_s^{v+h}) - a(v, X_s^v)\|_2^p \, ds \\ &\quad + C_5 \int_0^t E \, \|b(v+h, X_s^{v+h}) - b(v, X_s^v)\|^p \, ds \\ &\leq C_7 \|h\|^p + C_8 \int_0^t E \, \|X_s^{v+h} - X_s^v\|^p \, ds.\end{aligned}$$

Hence

$$\|X^{v+h} - X_v\|_{p,t}^p \leq C_7 \|h\|^p + C_8 \int_0^t \|X^{v+h} - X^v\|_{p,s}^p \, ds, \qquad t \in [0, 1].$$

Applying Gronwall's lemma now gives

$$\|X^{v+h} - X^v\|_{p,t}^p \leq C_9 \|h\|^p \qquad \text{for each } t \in [0, 1].$$

Inequality (3.9) now follows from this.

In order to prove that η defined in (3.8) is $(D_v X^v)h$, it must be shown that the remainder term $X^{v+h} - X^v - \eta$ is $o(h)$ in B_p as $h \to 0$ in $\mathbf{R}^d$. This is done as follows:

$$\begin{aligned}X_s^{v+h} - X_s^v - \eta_s &= x^{v+h} - x^v - Dx(v)h \\ &\quad + \int_0^t \{a(v, X_s^{v+h}) - a(v, X_s^v) - Da(v, X_s^v)(h, \eta_s)\} \, dw_s \\ &\quad + \int_0^t \{b(v, X_s^{v+h}) - b(v, X_s^v) - Db(v, X_s^v)(h, \eta_s)\} \, ds.\end{aligned}$$

Proceeding in the same way as before we obtain

$$\|X^{v+h} - X^v - \eta\|_{p,t}^p \le C_{10}\,\|x(v+h) - x(v) - Dx(v)h\|^p$$
$$+ C_{11}\int_0^t E\,\|a(v+h, X_s^{v+h}) - a(v, X_s^v)$$
$$- Da(v, X_s^v)(h, \eta_s)\|_2^p\,ds$$
$$+ C_{12}\int_0^t E\,\|b(v+h, X_s^{v+h}) - b(v, X_s^v)$$
$$- Db(v, X_s^v)(h, \eta_s)\|^p\,ds$$

Now apply Taylor's theorem to x at $v \in \mathbf{R}^d$ and to a and b at $(v, X_s^v) \in \mathbf{R}^{2d}$. The boundedness of the second derivatives of a and b, together with the Lipschitz property of X (in the norm $\|\ \|_{2p}$) imply that the second order remainder terms arising from a and b are $o(\|h\|^{2p})$. This yields the inequalities

$$\|X^{v+h} - X^v - \eta\|_{p,t}^p \le C_{11}\,\|h\|^{2p}$$
$$+ C_{13}\int_0^t E\{\|Da(v, X_s^v)(0, X_s^{v+h} - X_s^v - \eta_s\|_2^p$$
$$+ \|Db(v, X_s^v)(0, X_s^{v+h} - X_s^v - \eta_s)\|^p\}\,ds$$
$$\le C_{11}\,\|h\|^{2p} + C_{14}\int_0^t E\,\|X_s^{v+h} - X_s^v - \eta_s\|^p\,ds.$$

Hence

$$\|X^{v+h} - X^v - \eta\|_{p,t}^p$$
$$\le C_{11}\,\|h\|^{2p} + C_{15}\int_0^t E\,\|X^{v+h} - X^v - \eta\|_{p,s}^p\,ds$$

Another application of Gronwall's lemma now gives $\|X^{v+h} - X^v - \eta\|_{p,t}^p \le C_{16}\,\|h\|^{2p}$ for each $t \in [0, 1]$, and it follows that $X^{v+h} - X^v - \eta$ is $o(h)$ in B_p as $h \to 0$ in $\mathbf{R}^d$. Hence η is the derivative of X^v with respect to v in the direction h. ■

Theorem 3.2 Suppose that R, S and T are maps such that $R: N \to B_p(\mathbf{R}^d)$, $S: N \to B_p(L_2(\mathbf{R}^d, \mathbf{R}^n; \mathbf{R}^d))$†, $T: N \to B_p(L(\mathbf{R}^d; \mathbf{R}^d))$, for every $p \ge 1$. Furthermore suppose R, S and T are all C_∞^1 maps and S and T are bounded.

For each $v \in N$ let X^v denote the solution of the stochastic differential

† $L_2(\mathbf{R}^d, \mathbf{R}^n; \mathbf{R}^d)$ denotes the set of bilinear maps from $\mathbf{R}^d \times \mathbf{R}^n$ to $\mathbf{R}^d$.

equation

$$X_t^v = R(v)_t + \int_0^t S(v)_s(X_s^v, \mathrm{d}w_s) + \int_0^t T(v)_s X_s^v \,\mathrm{d}s, \qquad t \in [0, 1].$$

Then the map $v \to X^v$ is C^1_∞. For any $h \in \mathbf{R}^d$, $(D_v X^v)h$ is the solution η of the equation

$$\eta_t = DR(v)h_t + \int_0^t DS(v)h_s(X_s^v, \mathrm{d}w_s) + S(v)_s(\eta_s, \mathrm{d}w_s)$$

$$+ \int_0^t \{DT(v)h_s X_s^v + T(v)_s \eta_s\} \,\mathrm{d}s; \qquad t \in [0, 1].$$

We omit the proof of this result, simply remarking that it is similar to that of Theorem 3.1.

Note that if ϕ is a bounded, twice differentiable function on $\mathbf{R}^d$ with bounded first and second derivatives and $X \in C^1_\infty$, then $\phi(X)$ is a bounded process, also in C^1_∞. Furthermore the product of any two processes in C^1_∞ is again in C^1_∞. Hence combining Theorems 3.1 and 3.2 gives the following more general form of Theorem 3.1.

Theorem 3.3 *Suppose that the maps x, a and b in Theorem 3.1 are bounded with bounded derivatives up to order $m+1$. Then for every p, the map $v \to X^v$ is m times differentiable into B^p, with derivatives given by the solutions of the stochastic differential equations obtained by successive formal differentiation with respect to v in equation* (3.7).

3.3 Regularity of the measures induced by a stochastic differential equation

Assume throughout this section that A and B are bounded maps from $\mathbf{R}^d$ into $\mathbf{M}^{n,d}$ and $\mathbf{R}^d$ respectively with bounded derivatives of all orders. Let x be a point in $\mathbf{R}^d$ and ξ the solution of the stochastic differential equation

$$\xi_s = x + \int_0^s A(\xi_u)\,\mathrm{d}w_u + \int_0^s B(\xi_u)\,\mathrm{d}u, \qquad s \in [0, 1].$$

Let t be a fixed time in $(0, 1]$.

We will now describe the construction of the perturbing factor H in section 3.1 in the multi-dimensional case. Let I denote the $d \times d$ identity matrix and Y the solution of the following matrix-valued equation

$$Y_s = I + \int_0^s DA(\xi_u)Y_u \, dw_u + \int_0^s DB(\xi_u)Y_u \, du,\dagger \qquad s \in [0, 1] \qquad (3.10)$$

An application of Itô's lemma now shows that each of the matrices Y_t has an inverse, which we will denote by $\bar{Y}_t$, and that the process $\bar{Y}$ satisfies the equation

$$\bar{Y}_s = I - \int_0^s \bar{Y}_u DA(\xi_u) \, dw_u$$
$$+ \int_0^s \bar{Y}_u \left\{ -DB(\xi_u) + \sum_{i=1}^n DA(\xi_u)e_i DA(\xi_u)e_i \right\} du, \qquad s \in [0, 1].$$

Here $\{e_i\}_{i=1}^n$ is an orthonormal basis of $\mathbf{R}^n$. Note that Theorem 3.2 implies that both Y and $\bar{Y}$ are in B_p for all p.

For every $m \in \mathbf{N}$, define

$$H_s^m = \begin{cases} A(\xi_u)^* \bar{Y}_u^* & \text{if } \|A(\xi_u)^* \bar{Y}_u^*\| \le m \\ 0 & \text{otherwise} \end{cases}$$

where $\|\cdot\|$ denotes any (fixed) norm on $\mathbf{M}^{d,n}$.

In keeping with the notation in section 3.1, for every $v \in \mathbf{R}^d$ consider the perturbation of w defined by

$$W_s^{v,m} = w_s - \int_0^s H_u^m v \, du, \qquad s \in [0, 1]$$

and the corresponding stochastic differential equation

$$\xi_s^{v,m} = x + \int_0^s A(\xi_u^{v,m}) \, dw_u^{v,m} + \int_0^s B(\xi_u^{v,m}) \, du$$
$$= x + \int_0^s A(\xi_u^{v,m}) \, dw_u + \int_0^s \{B(\xi_u^{v,m}) - A(\xi_u^{v,m})H_u^m v\} \, du.$$

Let η^m be the matrix-valued process given by

$$\eta^m = \frac{\partial}{\partial v} \xi^{v,m}\Big|_{v=0}$$

in B^p for every p.

Note that this exists by the results of section 3.2 and satisfies the

† In this equation each $DA(\xi_u)$ is considered as a linear map from $\mathbf{R}^n$ to $\mathbf{M}^{d,d}$.

equation

$$\eta_s^m = I + \int_0^s DA(\xi_u)\eta_u^m \, dw_u + \int_0^s \{DB(\xi_u)\eta_u^m - A(\xi_u)H_u^m\} \, du$$

$$s \in [0, 1].$$

It follows from the method of variation of parameters that for any $s \in [0, 1]$

$$\eta_s^m = -Y_s \int_0^s \bar{Y}_u A(\xi_u) H_u^m \, du$$

where Y is as in equation (3.10).

For every s, define $\eta_s = \lim_{m\to\infty} \eta_s^m$. Then

$$\eta_s = -Y_s \int_0^s \bar{Y}_u A(\xi_u) A(\xi_u)_u^* \bar{Y}_u^* \quad \text{a.s.}$$

We now have the following result.

Theorem 3.4 *If the matrix η_t is non-degenerate with probability* 1 *and $\eta_t^{-1} \in L^p$ for every p, then the distribution of ξ_t is absolutely continuous with respect to Lebesgue measure on $\mathbf{R}^d$ and has a smooth density.*

Proof Define an operator L $(= L(m))$ by $L(\xi) = \eta^m$ where ξ and η^m are as above. Note that L can be applied to the solution of any stochastic differential equation driven by w whose coefficients satisfy the hypotheses of Theorem 3.2 when w is replaced by $w^{v,m}$. In particular L may be applied to ξ arbitrarily more times, and the result will always be a linear stochastic differential equation of the form:

$$(L^k\xi)_s = \int_0^s \{DA(\xi_u)(L^k\xi)_u + Q(u)\} \, dw_u$$

$$+ \int_0^s \{DB(\xi_u)(L^k\xi)_u + T(u)\} \, du$$

where both $Q(u)$ and $T(u)$ are products of bounded functions (involving the derivatives of A and B) and polynomial functions in the components of $(L\xi)_u$, $(L^2\xi)_u, \ldots, (L^{k-1}\xi)_u$, $(L\bar{Y})_u$, $(L^2\bar{Y})_u, \ldots, (L^{k-1}\bar{Y})_u$. It follows from Theorem 3.2 that each $L^k\xi$ is in B_p for every p. Note also that

$$\left.\begin{array}{l}\text{for each } k \text{ a bound can be found on } \|L^k\xi\|_p \\ \text{which is independent of } m\end{array}\right\} \qquad (3.11)$$

The same holds for the application of L to the stochastic integral

$$S_m h \equiv \int_0^t H_s^m h \, dw_s$$

where h is any vector in $\mathbf{R}^d$.

Suppose that for each ℓ, R_ℓ is a function on $\mathbf{M}^{d,d}$ with the following properties

(i) $0 \le R_\ell \le 1$.

(ii) $R_\ell(A) = \begin{cases} 1 & \text{if } A \text{ is invertible and } \|A^{-1}\| \le \ell \\ 0 & \text{if } \|A^{-1}\| \ge 2\ell \text{ or } A \text{ is degenerate.} \end{cases}$

(iii) R_ℓ has bounded derivatives of all orders.

Note that since, for each m, the process H^m is adapted and bounded, the Girsanov theorem applies to the perturbed process $w^{v,m}$. Denote by G_v^m the corresponding Girsanov density

$$G_v^m = \exp\left\{\int_0^1 H_u^m v \, dw_u - \frac{1}{2}\int_0^1 \|H_u^m v\|^2 \, du\right\}.$$

Suppose that ϕ, b and $y_1, y_2, \ldots, y_b$ are as in Lemma 1.14. Then for each i and j the quantity below is independent of v.

$$E[D^{b-1}\phi(\xi_t^v)(y_1, \ldots, y_{b-1})\bar{\eta}_{t_{ij}}^{v,m} R_\ell(\eta_t^{v,m})G_v^m]. \tag{3.12}$$

Let $x_1, x_2, \ldots, x_d$ be an orthonormal basis of $\mathbf{R}^d$. Differentiating in (3.12) with respect to v in the direction x_i, then setting v to 0 and equating the resulting expression to zero gives

$$\begin{aligned} &E[D^b\phi(\xi_t)(y_1, \ldots, y_{b-1}, \eta_t x_i)\bar{\eta}_{t_{ij}}^m R_\ell(\eta_t^m)] \\ &\quad = -E\left[D^{b-1}\phi(\xi_t)(y_1, \ldots, y_{b-1})\frac{\partial}{\partial v}\{\overline{\eta_{t_{ij}}^{v,m}}\, R_\ell(\eta_t^{v,m})G_v^m\}x_i|_{v=0}\right] \end{aligned} \tag{3.13}$$

The expression

$$\frac{\partial}{\partial v}\{\overline{\eta_{t_{ij}}^{v,m}}\, R_\ell(\eta_t^{v,m})G_v^m\}x_i|_{v=0}$$

is a polynomial in the components of $\bar{\eta}_t^m$, $(L\xi)_t$, $(L^2\xi)_t$, $R_\ell(\eta_t^m)$, $R'_\ell(\eta_t^m)$ and $S_m x_i$. If

$$y_b = \sum_{j=1}^d y^j x_j$$

then

$$\sum_{i,j=1}^d \bar{\eta}_{t_{ij}}^m y^j x_i = \bar{\eta}_t^m y_b$$

so multiplying each side of (3.13) by y^j and summing over i and j will give

$$E[D^b\phi(\xi_t)(y_1, \ldots, y_{b-1}, \eta_t\bar{\eta}_t^m y_b)R_\ell(\eta_t^m)]$$
$$= -E[D^{b-1}\phi(\xi_t)(y_1, \ldots, y_{b-1})P_1^m]$$

where P_1^m is a polynomial in the terms described above.

By successively iterating this procedure in the fashion described in section 3.1, we arrive at the following system of equalities.

$$E[D^b\phi(\xi_t)(y_1, \ldots, y_{b-1}, \eta_t\bar{\eta}_t^m y_b)R_\ell(\eta_t^m)]$$
$$= -E[D^{b-1}\phi(\xi_t)(y_1, \ldots, y_{b-1})P_1^m]$$
$$E[D^{b-1}\phi(\xi_t)(y_1, \ldots, y_{b-2}, \eta_t\bar{\eta}_t^m y_{b-1})P_1^m]$$
$$= -E[D^{b-2}\phi(\xi_t)(y_1, \ldots, y_{b-2})P_2^m]$$
$$\vdots$$
$$E[D\phi(\xi_t)(\eta_t\bar{\eta}_t^m y_1)P_{b-1}^m] = -E[\phi(\xi_t)P_b^m].$$

Here each P_j^m, $1 \le j \le b$, is a polynomial in the components of $\bar{\eta}_t^m$, $(L\xi)_t, \ldots, (L^{j+1}\xi)_t$, $R_\ell(\eta_t^m)$, $R'_\ell(\eta_t^m), \ldots, R_\ell^{(j)}(\eta_t^m)$ and $S_m x_i$, $L(S_m x_i), \ldots, L^{j-1}(S_m x_i)$.

Now take the limit as $m \to \infty$. The conditions on R_ℓ, together with (3.11), imply that the L^1-norm of each polynomial P_j^m is bounded in m. This gives

$$|E[D^b\phi(\xi_t)(y_1, \ldots, y_{b-1}, y_b)R_\ell(\eta_t)]| \le \|\phi\|_\infty Q_\ell \tag{3.14}$$

where Q_ℓ is a product of L_p norms (for various p) of $\bar{\eta}_t$, $R_\ell(\eta_t), \ldots, R_\ell^{(j)}(\eta_t)$.

Finally, the hypothesis that $\bar{\eta}_t \in L^p$ for all p implies that Q_ℓ is bounded in ℓ, so taking the limit as $\ell \to \infty$ yields inequality (1.8). Hence it follows that ξ_t has a smooth density. ∎

4

An elementary derivation of Malliavin's inequalities

4.1 Introduction

In this chapter we will rederive the main theorems of chapters 2 and 3 by means of an elementary argument. This work first appeared in Bell[1]. Our approach is outlined below and presented in detail in sections 4.2–4.4.

Consider first an analogous finite dimensional problem. Let $d\gamma_m(x) = (2\pi)^{-m/2}\exp(-\|x\|^2/2)\,dx$ denote the standard normal measure on $\mathbf{R}^m$ and let F be a C^2 map from $\mathbf{R}^m$ into $\mathbf{R}^d$ with a surjective derivative $DF(x)$ at each point x. Suppose that one wishes to study the regularity of the induced measure $\nu = F(\gamma_m)$ on $\mathbf{R}^d$. Then a natural way to proceed is as follows. Given a vector $y \in \mathbf{R}^d$, define a C^1 function h on $\mathbf{R}^m$ such that $DF(x)h(x) \equiv y$. For any test function ϕ on $\mathbf{R}^d$, integration by parts gives

$$\int_{\mathbf{R}^d} D\phi(z)y\,d\nu(z) = \int_{\mathbf{R}^m} D(\phi \circ F)(x)h(x)\,d\gamma_m(x)$$

$$= \int_{\mathbf{R}^m} \phi \circ F(x)X(x)\,d\gamma_m(x)$$

where $X(x) = \langle h(x), x\rangle - \text{Trace}\,Dh(x)$. Assume that for each y the corresponding function X is integrable with respect to γ_m. We will then have

$$\left|\int_{\mathbf{R}^d} D\phi(z)y\,d\nu(z)\right| \le \|X\|_{L^1(\gamma_m)}\,\|\phi\|_\infty .$$

It follows from this and Lemma 1.12 that ν is absolutely continuous with respect to the Lebesgue measure on $\mathbf{R}^d$.

Suppose now that ν is the measure induced on $\mathbf{R}^d$ by the random variable ξ_t where, as before, ξ is the solution of a stochastic differential equation driven by n-dimensional Brownian motion w, and t is a fixed positive time. Then ν is the image of the n-dimensional Wiener measure γ under the Itô map $g: w \to \xi$, composed with evaluation at t. Let us denote by g_t the map defined by $w \to \xi_t$. In Definition 4.2 we introduce a sequence of differentiable functions $\{g^m : m = 1, 2, \ldots\}$ on C_0. We show in Theorem 4.6 that $g^m(w) \to g(w)$ in $L^2(\gamma)$. The absolute continuity of γ under suitable assumptions is proved in Theorem 4.9.

In keeping with the argument above, the proof involves constructing for each $y \in \mathbf{R}^d$, a sequence of path-valued maps $\{h^m\}$ on C_0 such that

$$Dg_t^m(w)h^m(w) \equiv y.\dagger$$

The sequences $\{g^m\}$ and $\{h^m\}$ are constructed so that they factor through finite dimensional subspaces $\{\mathbf{V}_m\}$ of C_0. For each m, let P_m denote the projection from C_0 onto $\mathbf{V}_m$ and note that the measure $P_m(\gamma)$ on $\mathbf{V}_m$ is normal. By applying the dominated convergence theorem and integrating by parts‡ with respect to the measures $P_m(\gamma)$ we obtain

$$\begin{aligned}
\int_{\mathbf{R}^d} D\phi(z)y \, d\nu(z) &= \lim_{m\to\infty} \int_{C_0} D\phi(g_t^m(w))y \, d\gamma(w) \\
&= \lim_{m\to\infty} \int_{C_0} D(\phi \circ g_t^m)(w)h^m(w) \, d\gamma(w) \\
&= \lim_{m\to\infty} \int_{\mathbf{V}_m} D(\phi \circ g_t^m)(x)h^m(x) \, dP_m(\gamma)(x) \\
&= \lim_{m\to\infty} \int_{\mathbf{V}_m} \phi \circ g_t^m(x)\{\langle h^m(x), x\rangle_m \\
&\qquad - \mathrm{Trace}_m \, Dh^m(x)\} \, dP_m(\gamma)(x)
\end{aligned}$$

Here $\langle , \rangle_m$ denotes the inner product induced on $\mathbf{V}_m$ by P_m. Writing these as integrals with respect to the original measure space (C_0, γ) gives the relation

$$\int_{\mathbf{R}^d} D\phi(z)h \, d\nu(z) = \lim_{m\to\infty} \int_{C_0} \phi \circ g_t^m(w)X_m(w) \, d\gamma(w)$$

where $X_m(w) = \langle h^m(w), P_m w\rangle_{L_0^{2,1}} - \mathrm{Trace}_{L_0^{2,1}} Dh^m(w)$.

Lemmas 4.1, 4.4 and 4.5 are used to show that the sequence $\{X_m : m = 1, 2, \ldots\}$ is bounded in $L^1(\gamma)$. The above relation then implies that

$$\left| \int_{\mathbf{R}^d} D\phi(z)y \, d\nu(z) \right| \leq \sup_m \|X_m\|_{L^1(\gamma)} \|\phi\|_\infty$$

and the absolute continuity of ν follows as before.

† It is possible to construct h^m in this way only for sufficiently large values of m. This necessitates the use of the 'cut-off functions' R_N in section 4.4. The idea of introducing functions of this kind comes from Malliavin's original paper.

‡ This is equivalent to the application of the divergence theorem in the proof of Theorem 4.9.

In Theorem 4.10 we obtain the estimates required to prove smoothness of the density of ν. This is done by iterating the integration by parts operation in the proof of Theorem 4.9.

In the sequel, $\{x_1, x_2, \ldots, x_n\}$ and $\{z_1, z_2, \ldots, z_d\}$ will denote fixed orthonormal bases of $\mathbf{R}^n$ and $\mathbf{R}^d$ respectively. If H is a bilinear map from $\mathbf{R}^n \times \mathbf{R}^n$ into $\mathbf{R}$, then $\operatorname{Tr} H$ will denote the trace of H, defined by

$$\operatorname{Tr} H = \sum_{i=1}^{n} H(x_i, x_i).$$

The ℓ^2-norm on $\mathbf{R}^d$ will be denoted throughout by $\|\ \|$. Thus if K is a linear map from $\mathbf{R}^n$ to $\mathbf{R}^d$ then

$$\operatorname{Tr} \|K\|^2 = \sum_{i=1}^{n} \|Kx_i\|^2.$$

Finally note that the bounds $d_1, d_2, \ldots, d_{21}$ appearing in the proofs of the theorems are all independent of the parameter m above.

4.2 A sequence of differentiable approximations to the Itô map

The following result is a discrete version of Gronwall's lemma.

Lemma 4.1 *Suppose that for some $m \in \mathbf{N}$ and positive constants C and D, numbers $\lambda_0, \lambda_1, \ldots, \lambda_m$ satisfy the inequalities*

$$\lambda_0 \leq C$$

$$\lambda_k \leq C + \frac{D}{m} \sum_{j=0}^{k-1} \lambda_j; \quad k = 1, 2, \ldots, m.$$

Then

$$\lambda_k \leq C(1 + D/m)^k \quad \textit{for all} \quad 0 \leq k \leq m. \tag{4.1}$$

In particular $\lambda_k \leq Ce^D$ for all $k = 0, 1, \ldots, m$.

Proof The required inequality holds for $k = 0$. Assume that (4.1) holds for $k = 1, 2, \ldots, i$. Then

$$\begin{aligned}
\lambda_{i+1} &\leq C + \frac{D}{m} \sum_{j=0}^{i} \lambda_j \\
&\leq C + \frac{CD}{m} \sum_{j=0}^{i} \left(1 + \frac{D}{m}\right)^j \\
&= C + \frac{CD}{m} \left\{ \frac{(1 + D/m)^{i+1} - 1}{(1 + D/m) - 1} \right\} = C\left(1 + \frac{D}{m}\right)^{i+1}.
\end{aligned}$$

Hence (4.1) follows by induction on k. Furthermore since

$$C\left(1+\frac{D}{m}\right)^k \leq C\left(1+\frac{D}{m}\right)^m \uparrow Ce^D \quad \text{as} \quad m\to\infty$$

it follows that $\{\lambda_k\}_{k=0}^m$ is bounded as required. ■

Notation Throughout this chapter let t denote a fixed element of $(0, 1]$ and for $m\in\mathbf{N}$ and $w\in C_0$ denote by $\Delta_j w$ $(=\Delta_j^m w)$ the increment $w((j+1)t/m)-w(jt/m)$; $j=0, 1, \ldots, m-1$. Assume further that x is a point in $\mathbf{R}^d$ and that A and B are maps from $\mathbf{R}^d$ into $\mathbf{M}^{n,d}$ and $\mathbf{R}^d$ respectively.

Definition 4.2 For $m\in\mathbf{N}$ and each path $w\in C_0$ define elements $v_0, v_{t/m}, \ldots, v_t$ in $\mathbf{R}^d$ inductively as follows.

$$v_0 = x$$

$$v_{kt/m} = x + \sum_{j=0}^{k-1} A(v_{jt/m})\Delta_j w + \frac{t}{m}\sum_{j=0}^{k-1} B(v_{jt/m}); \qquad k=1,\ldots,m \qquad (4.2)$$

Let v $(=v^m)$ denote the path in $L^{2,1}$ which is piecewise linear between the points $0, t/m, \ldots, t$ and constant on $[t, 1]$. Finally, define g^m from C_0 into $L^{2,1}$ to be the map which carries w to v.

Note that $v_0, v_{t/m}, \ldots, v_t$ in (4.2) are measurable with respect to the σ-fields $\mathcal{F}_0, \mathcal{F}_{t/m}, \ldots, \mathcal{F}_t$ respectively, where $\{\mathcal{F}_s : 0\leq s\leq 1\}$ is the filtration for w.

Theorem 4.3 *Suppose that A and B are C^r functions. Then for each m, the map g^m is also C^r.*

The proof is an easy inductive argument on k in (4.2).

The inequalities contained in the following lemma will be needed later on. Their proofs are simple applications of the triangle and Holder inequalities.

Lemma 4.4 (i) *Suppose that $y_1, y_2, \ldots, y_r$ are in $\mathbf{R}^d$. Then for every $p\in\mathbf{N}$*

$$\left\|\frac{1}{r}\sum_{i=1}^r y_i\right\|^p \leq \frac{1}{r}\sum_{i=1}^r \|y_i\|^p.$$

In particular

$$\left\|\sum_{i=1}^{r} y_i\right\|^p \leq C \sum_{i=1}^{r} \|y_i\|^p$$

where C depends only upon r and p.
(ii) *Suppose that* $X_1, X_2, \ldots, X_r$ *are* L^p *random variables with values in* $\mathbf{R}^d$. *Then*

$$E\left\|\frac{1}{r}\sum_{i=1}^{r} X_i\right\|^p \leq \max\{E\|X_i\|^{2^{p-1}} : 1 \leq i \leq r\}.$$

(iii) *Suppose that p is a positive integral power of 2 and* $f:[0,1]\to\mathbf{R}^d$. *Then*

$$\left\|\int_0^1 f(s)\,\mathrm{d}s\right\|^p \leq \int_0^1 \|f(s)\|^p\,\mathrm{d}s$$

provided the right-hand side exists.

Lemma 4.5 *Suppose that* $X, V_0, \ldots, V_{m-1}$ *are* $\mathbf{R}^d$*-valued random variables;* $U_0, \ldots, U_{m-1}$ *and* $Z_0, \ldots, Z_{m-1}$ *are random linear maps from* $\mathbf{R}^n$ *to* $\mathbf{R}^d$ *and from* $\mathbf{R}^d$ *to* $\mathbf{R}^d$ *respectively; and that* $Y_0, \ldots, Y_{m-1}$ *are random bilinear maps from* $\mathbf{R}^d \times \mathbf{R}^n$ *to* $\mathbf{R}^d$. *Assume that these satisfy the following conditions.*
(i) U_i, V_i, Y_i *and* Z_i *are measurable with respect to* $\mathcal{F}_{it/m}$ *for each* $i = 0, 1, \ldots, m-1$.
(ii) *For some p of the form* 2^r *where* $r \in \mathbf{N}$, *there exists a constant M such that*

$$\max\{\|X\|_{L^p}; \|U_i\|_{L^p}, \|V_i\|_{L^p}, \|Y_i\|_{L^\infty}, \|Z_i\|_{L^\infty}; \quad 0 \leq i \leq m\} \leq \mathrm{M}.$$

Suppose $\eta_0, \eta_{t/m}, \ldots, \eta_t$ *satisfy the equations*

$$\begin{aligned}\eta_{kt/m} = x &+ \sum_{j=0}^{k-1} U_j(\Delta_j w) + \frac{1}{m}\sum_{j=0}^{k-1} V_j \\ &+ \sum_{j=0}^{k-1} Y_j(\eta_{jt/m}, \Delta_j w) \\ &+ \frac{1}{m}\sum_{j=0}^{k-1} Z_j(\eta_{jt/m}).\end{aligned}$$

Then $E\|\eta_{kt/m}\|^p \leq N$ *for each* $k = 0, \ldots, m$; *where N is a constant which depends only on M and p.*

Proof Using Lemma 4.4(i) we have

$$\|\eta_{kt/m}\|^p \le d_1\Big[\|X\|^p + \Big\|\sum_{j=0}^{k-1} U_j(\Delta_j w)\Big\|^p + \Big\|\frac{1}{m}\sum_{j=0}^{k-1} V_j\Big\|^p$$
$$+ \Big\|\sum_{j=0}^{k-1} Y_j(\eta_{jt/m}, \Delta_j w)\Big\|^p + \Big\|\frac{1}{m}\sum_{j=0}^{k-1} Z_j(\eta_{jt/m})\Big\|^p\Big]$$

The second and fourth summations on the right-hand side are stochastic integrals, so their expectations may be estimated by Theorem 1.9(iv). Together with Lemma 4.4(i) this gives

$$E\,\|\eta_{kt/m}\|^p \le d_1\,\|X\|^p + \frac{d_2}{m}\sum_{j=0}^{k-1} E[\mathrm{Tr}\,\|U_j\|^2]^{p/2} + \frac{d_3}{m}\sum_{j=0}^{k-1} E\,\|V_j\|^p$$
$$+\frac{d_4}{m}\sum_{j=0}^{k-1} E[\mathrm{Tr}\,\|Y_j(\eta_{jt/m}, \cdot)\|^2]^{p/2}$$
$$+\frac{d_5}{m}\sum_{j=0}^{k-1} E\,\|Z_j\eta_{jt/m}\|^p.$$

Using (ii) we have

$$E\,\|\eta_{kt/m}\|^p \le d_6 + \frac{d_2}{m}\sum_{j=0}^{k-1} E\,\|U_j\|^p + \frac{d_3}{m}\sum_{j=0}^{k-1} E\,\|V_j\|^p$$
$$+\frac{d_4}{m}\sum_{j=0}^{k-1} E[\|Y_j\|^p\,\|\eta_{jt/m}\|^p]$$
$$+\frac{d_5}{m}\sum_{j=0}^{k-1} E\,[\|Z_j\|^p\,\|\eta_{jt/m}\|^p]$$
$$\le d_7 + \frac{d_8}{m}\sum_{j=0}^{k-1} E\,\|\eta_{jt/m}\|^p$$

where the above constants d_7 and d_8 depend only upon M and p. The result now follows from Lemma 4.1 by defining each λ_k as $E\,\|\eta_{kt/m}\|^p$. ∎

Theorem 4.6 *Suppose that A and B satisfy the following conditions*
(i) *A and B are C^2.*
(ii) *A and B, together with their first and second order derivatives, are bounded.*
Then

$$\sup_{s\in[0,t]} E\,\|\xi_s - v_s\|^p \to 0$$

as m tends to infinity for every integer $p \in \mathbf{N}$, where v is as in

Definition 4.2 *and* ξ *is the solution of the stochastic differential equation*

$$\xi_s = x + \int_0^s A(\xi_u)\, dw_u + \int_0^s B(\xi_u)\, du; \qquad s \in [0, 1].$$

Proof Note that for any $0 \le k \le m-1$ and $kt/m \le s \le (k+1)t/m$

$$\xi_s - v_{kt/m} = \sum_{j=0}^{k-1} \int_{jt/m}^{(j+1)t/m} [A(\xi_u) - A(v_{jt/m})]\, dw_u$$
$$+ \sum_{j=0}^{k-1} \int_{jt/m}^{(j+1)t/m} [B(\xi_u - B(v_{jt/m})]\, du$$
$$+ \int_{k/m}^{s} A(\xi_u)\, dw_u + \int_{k/m}^{s} B(\xi_u)\, du.$$

Denoting $A(\xi_u) - A(v_{jt/m})$ by $f(u)$ and $B(\xi_u) - B(v_{jt/m})$ by $g(u)$ for $u \in (jt/m, (j+1)t/m)$, we have

$$\xi_s - v_{kt/m} = \int_0^{kt/m} f(u)\, dw_u + \int_0^{kt/m} g(u)\, du$$
$$+ \int_{kt/m}^{s} A(\xi_u)\, dw_u + \int_{kt/m}^{s} B(\xi_u)\, du.$$

It is sufficient to prove the theorem for every p of the form 2^r where $r \in \mathbf{N}$. Suppose that p is of this form. Then estimating the L^p-norm of the right-hand side in the above expression by means of Lemma 4.4 and Theorem 1.9(iv) gives

$$E\,\|\xi_s - v_{kt/m}\|^p \le d_9 \int_0^{kt/m} \{[E\,\mathrm{Tr}\,\|f(u)\|^2]^{p/2} + E\,\|g(u)\|^p\}\, du$$
$$+ d_{10} \int_{kt/m}^{s} \{[E\,\mathrm{Tr}\,\|A(\xi_u)\|^2]^{p/2} + E\,\|B(\xi_u)\|^p\}\, du.$$

Condition (ii) implies that A and B are globally Lipschitz maps. Using this, together with the boundedness of A and B, we obtain

$$E\,\|\xi_s - v_{kt/m}\|^p \le \frac{d_{11}}{m} + d_{12} \sum_{j=0}^{k-1} \int_{jt/m}^{(j+1)t/m} E\,\|\xi_u - v_{jt/m}\|^p\, ds$$

Thus

$$\lambda_k \le \frac{d_{11}}{m} + \frac{d_{12}}{m} \sum_{j=0}^{k-1} \lambda_j; \qquad k = 0, 1, \ldots, m-1$$

where each λ_j denotes $\sup\{E\|\xi_s - v_{jt/m}\|^p : jt/m \le s \le (j+1)t/m\}$. Lemma 4.1 now gives

$$\lambda_k \le d_{11}e^{d_{12}}/m; \qquad k = 0, 1, \ldots, m-1. \tag{4.3}$$

In particular

$$E\|\xi_{kt/m} - v_{kt/m}\|^p \le d_{13}/m \quad \text{for} \quad k = 0, 1, \ldots, m-1 \tag{4.4}$$

Using Theorem 1.9(iv) one obtains the estimates

$$E\|\xi_s - \xi_{kt/m}\|^p \le d_{14}/m \quad \text{for}$$
$$s \in [kt/m, (k+1)t/m]; \qquad k = 0, 1, \ldots, m-1. \tag{4.5}$$

Furthermore for $s \in [kt/m, (k+1)t/m]$,

$$\|v_s - v_{kt/m}\| \le \|v_{(k+1)t/m} - v_{kt/m}\|$$

and this implies

$$E\|v_s - v_{kt/m}\|^p \le d_{15}/m \quad \text{for}$$
$$s \in [kt/m, (k+1)t/m]; \qquad k = 0, 1, \ldots, m-1. \tag{4.6}$$

The theorem now follows from (4.4), (4.5), (4.6) and the triangle inequality. ■

4.3 Derivation of the covariance matrix

Suppose throughout this section that A and B are C^2, bounded, and have bounded first and second order derivatives. Note that this implies that A, B, DA and DB are all globally Lipschitz maps. Let w denote a typical realization of n-dimensional Brownian motion and for each $m \in \mathbf{N}$ define a $d \times d$ matrix process σ_m by

$$\sigma^m(w) = Dg_t^m(w)Dg_t^m(w)^*$$

where * denotes the adjoint of a linear map between the Hilbert spaces $L_0^{2,1}$ and $\mathbf{R}^d$. In this section we will show that σ^m converges in probability to a limit σ as m tends to infinity. We will make use of the following easily verifiable result.

Lemma 4.7 *Suppose that f and g are random functions from $[0, t]$ to $\mathbf{R}$ and $\{a_{km} : k = 0, \ldots, m-1;\ m = 1, 2, \ldots\}$ and $\{b_{km} : k = 0, \ldots, m-1;\ m = 1, 2, \ldots\}$ are random sequences such that*

(i) *The functions $s \to E[f^2(s)]$ and $s \to E[g^2(s)]$ are bounded on $[0, t]$.*

(ii)

$$\max_{k=0,\ldots,m-1,s\in[kt/m,(k+1)t/m]} \sup E[a_{km}-f(s)]^2 \to 0 \quad \text{as} \quad m\to\infty$$

$$\max_{k=0,\ldots,m-1,s\in[kt/m,(k+1)t/m]} \sup E[b_{km}-g(s)]^2 \to 0 \quad \text{as} \quad m\to\infty.$$

Then

$$\frac{t}{m}\sum_{j=0}^{m-1} a_{km}b_{km} \to \int_0^t f(s)g(s)\,\mathrm{d}s$$

in L^1 as $m\to\infty$.

Theorem 4.8 *The matrix sequence $\{\sigma^m : m=1,2,\ldots\}$ converges in probability to a limit $\sigma=(\sigma_{ij})_{i,j=1}^d$, where*

$$\sigma_{ij}=\sum_{r=0}^{n}\int_0^t \langle z_i, \tilde{\eta}_t^{rs}\rangle\langle z_j, \tilde{\eta}_t^{rs}\rangle\,\mathrm{d}s; \qquad 1\le i,j\le d \tag{4.7}$$

and for each $s\in[0,t]$, $\tilde{\eta}^{rs}$ is defined by the equation

$$\tilde{\eta}_u^{rs}=0 \qquad \text{if } u\le s$$

$$\tilde{\eta}_u^{rs}=A(\xi_s)x_r+\int_s^u DA(\xi_v)(\tilde{\eta}_v^{rs},\mathrm{d}w_v) + \int_s^u DB(\xi_v)\tilde{\eta}_v^{rs}\,\mathrm{d}v \qquad \text{if } s<u\le t \tag{4.8}$$

Proof Note that for $1\le i,j\le d$

$$\sigma_{ij}^m=\sum_{n=1}^{\infty}\langle z_i, Dg_t^m(w)f_n\rangle\langle z_j, Dg_t^m(w)f_n\rangle \tag{4.9}$$

where $\{f_n : n=1,2,\ldots\}$ is any orthonormal basis of $L_0^{2,1}$.

For each $r\in\{1,2,\ldots,n\}$ and $k\in\{0,1,\ldots,m-1\}$ define a vector $f^{rk}\in L_0^{2,1}$ by

$$f_s^{rk}=0 \qquad \text{if } 0\le s\le tk/m$$
$$\sqrt{(m/t)}(s-kt/m)x_r \qquad \text{if } kt/m<s<(k+1)t/m \tag{4.10}$$
$$\sqrt{(t/m)}x_r \qquad \text{if } (k+1)t/m\le s\le 1$$

Then the set $S\equiv\{f^{rk} : r=1,\ldots,n;\ k=0,\ldots,m-1\}$ is orthonormal in $L_0^{2,1}$. Extend S to an orthonormal basis B_m of $L_0^{2,1}$. Note for any $f\in B_m\backslash S$ we will have $Dg_t^m(w)f=0$.

Evaluating the summation in (4.9) with respect to the basis B_m gives

$$\sigma_{ij}^m = \sum_{r=1,k=0}^{n,m-1} \langle z_i, Dg_t^m(w)f^{rk}\rangle \langle z_j, Dg_t^m(w)f^{rk}\rangle$$

$$= \frac{t}{m} \sum_{r=1,k=0}^{n,m-1} \langle z_i, \eta_t^{rk}\rangle \langle z_j, \eta_t^{rk}\rangle$$

where $\eta^{rk} = \sqrt{(m/t)}Dg^m(w)f^{rk}$. In view of Lemma 4.7 it suffices to prove that for each $1 \leq r \leq n$

$$\sup_{s\in[0,t]} E\,\|\tilde{\eta}_t^{rs}\|^2 < \infty \tag{4.11}$$

and

$$\max\{\sup_{kt/m\leq s\leq(k+1)t/m} E\,\|\tilde{\eta}_t^{rs} - \eta_t^{rk}\|^2 : k = 0, \ldots, m-1\} \to 0$$

$$\text{as} \quad m \to \infty. \tag{4.12}$$

These will be established by an argument similar to the proof of Theorem 4.6. Using Lemma 4.4 and Theorem 1.9(iv) to estimate the L^2-norm of the right-hand side in (4.8), then applying Gronwall's lemma, gives

$$\sup\{E\,\|\tilde{\eta}_u^{rs}\|^4 : 0 \leq s, u \leq t\} < \infty \tag{4.13}$$

In particular this implies (4.11) above. Differentiating in (4.2), we have

$$\eta_{jt/m}^{rk} = 0 \qquad \text{if } j \leq m$$

$$\eta_{jt/m}^{rk} = A(v_{kt/m})x_r + \sum_{i=k+1}^{j-1} DA(v_{it/m})(\eta_{it/m}^{rk}, \Delta_i w)$$

$$+ \frac{t}{m} \sum_{i=k+1}^{j-1} DB(v_{it/m})\eta_{it/m}^{rk} \qquad \text{if } k+1 \leq j \leq m.\dagger \tag{4.14}$$

Suppose that $j > k$, $kt/m \leq s \leq (k+1)t/m$, $jt/m \leq u \leq (j+1)t/m$, and write $\tilde{\eta}_u^{rs} - \eta_{jt/m}^{rk}$ in the form

$$\tilde{\eta}_u^{rs} - \eta_{jt/m}^{rk} =$$

$$X + \sum_{i=k+1}^{j-1} \left\{ \int_{it/m}^{(i+1)t/m} [DA(\xi_v)(\tilde{\eta}_v^{rs}, \cdot) - DA(v_{it/m})(\eta_{it/m}^{rk}, \cdot)]\, dw_v \right.$$

$$\left. + \int_{it/m}^{(i+1)t/m} [DB(\xi_v)\tilde{\eta}_v^{rs} - DB(v_{it/m})\eta_{it/m}^{rk}]\, dv \right\} \tag{4.15}$$

† The summations are defined to be zero if $j = k+1$.

It follows from (4.11) that $E\,\|X\|^2 \le d_{16}/m$. Applying Lemma 4.4 and Theorem 1.9(iv) to (4.15) and using the bounded and Lipschitz properties of DA and DB, together with (4.6), gives

$$E\,\|\tilde{\eta}_u^{rs} - \eta_{jt/m}^{rk}\|^2 \le \frac{d_{17}}{m}\left\{1 + \sum_{i=k+1}^{j-1} \sup_{v \in [it/m,(i+1)t/m]} E\,\|\tilde{\eta}_v^{rs} - \eta_{it/m}^{rk}\|^2\right\}$$

It now follows from Lemma 4.1 that

$$\sup_{u \in [jt/m,(j+1)t/m]} E\,\|\tilde{\eta}_u^{rs} - \eta_{jt/m}^{rk}\|^2 \le d_{18}/m$$

where the constant d_{18} depends only upon the maps A and B. This implies (4.12) and completes the proof of the theorem. ■

4.4 Regularity of the measures induced by a stochastic differential equation

Let $\mathbf{M}^d$ denote the ring of $d \times d$ matrices endowed with the Hilbert–Schmidt norm $\|\ \|$.

We will construct a sequence of smooth bump functions $\{R_N\}_{N=1}^{\infty}$ on $\mathbf{M}^d$ as follows: Let $\{\psi_N\}_{N=1}^{\infty}$ be a sequence of smooth functions from $[0, \infty)$ to $[0, 1]$ such that

(i) For each N, $\psi_N(t) = 1$ if $0 \le t \le N^2$; $\psi_N(t) = 0$ if $t \ge (N+1)^2$.

(ii) For every $p \in \mathbf{N}$, $\sup_{N,t} |D^p \psi_N(t)| < \infty$. Define

$$\begin{cases} R_N(\alpha) = \psi_N(\|\alpha^{-1}\|^2) & \text{if } \alpha \in GL(d) \\ 0 & \text{if } \alpha \in \mathbf{M}^d \backslash GL(d). \end{cases}$$

Since $GL(d)$ is an open subset in $\mathbf{M}^d$, it follows that each R_N is a C^∞ function. Furthermore, one can show that for every $p \in \mathbf{N}$ there exists $q \in \mathbf{N}$ and a constant C_p such that

$$\sup_N \|D^p R_N(\alpha)\| \le C_p\,\|\alpha^{-1}\|^q \tag{4.16}$$

for $\alpha \in GL(d)$.

The sequence $\{R_N\}$ will be used in the proofs of Theorems 4.9 and 4.10.

Theorem 4.9 *Suppose that $x \in \mathbf{R}^d$ and the maps $A: \mathbf{R}^d \to \mathbf{M}^{n,d}$ and $B: \mathbf{R}^d \to \mathbf{R}^d$ are both C^2, bounded and have bounded first and second order derivatives. Let ξ be the solution of the stochastic differential* *equation*

$$\xi_s = x + \int_0^s A(\xi_u)\,dw_u + \int_0^s B(\xi_u)\,du, \qquad s \in [0, 1].$$

Suppose that the matrix σ, defined in (4.7), *is invertible with probability* 1. *Then ξ_t has an absolutely continuous distribution with respect to Lebesgue measure on $\mathbf{R}^d$.*

Proof Let v denote the measure induced by ξ_t on $\mathbf{R}^d$. Define sequences of measures $\{\gamma_N\}_{N\geq 1}$ and $\{v_N\}_{N\geq 1}$ on C_0 and $\mathbf{R}^d$ respectively by

$$\frac{d\gamma_N}{d\gamma}(w) = R_N(\sigma(w))$$

$$v_N = g_t(\gamma_N).$$

The non-degeneracy assumption on σ implies that v_N converges to v in variation, so it suffices to prove that v_N is absolutely continuous for each fixed N. Let $y \in \mathbf{R}^d$ and for each $m \in \mathbf{N}$ define $h^m : C_0 \to L_0^{2,1}$ by

$$\begin{cases} h^m(w) = Dg_t^m(w)^*[\sigma^m(w)]^{-1}y & \text{if } \sigma^m \in GL(d) \\ 0 & \text{otherwise.} \end{cases}$$

Let ϕ be a real-valued C^∞ function on $\mathbf{R}^d$ with compact support and note that

$$D\phi(g_t^m(w))yR_N(\sigma^m) \equiv D[\phi \circ g_t^m](w)h^m R_N(\sigma^m).\dagger \tag{4.17}$$

Using the dominated convergence theorem together with (4.17) and then applying Theorem 1.4 one obtains

$$\begin{aligned} \int_{\mathbf{R}^d} D\phi(z)y\,dv_N(z) &= \int_{C_0} D\phi(g_t(w))yR_N(\sigma)\,d\gamma(w) \\ &= \lim_{m\to\infty} \int_{C_0} D(\phi \circ g_t^m)(w)h^m R_N(\sigma^m)\,d\gamma(w) \\ &= \lim_{m\to\infty} \int_{C_0} \phi \circ g_t^m(w)\mathrm{Div}[h^m R_N(\sigma^m)]\,d\gamma(w) \end{aligned} \tag{4.18}$$

where Div denotes the divergence operator defined in (1.2). We will show that

$$C \equiv \sup_m \|\mathrm{Div}[h^m R_N(\sigma^m)]\|_{L^1(\gamma)} < \infty. \tag{4.19}$$

† h^m is actually the smallest vector in $L_0^{2,1}$ satisfying this condition.

Since we have

$$\left|\int_{\mathbf{R}^d} D\phi(z)y \, \mathrm{d}\nu_N(z)\right| \le C \, \|\phi\|_\infty$$

the absolute continuity of ν_N will then follow from Lemma 1.12.

First consider the inner product term in Div. Note that this is non-zero only if $\sigma_m \in GL(d)$ and $\|(\sigma^m)^{-1}\| \le (N+1)$, in which case

$$\langle h^m R_N(\sigma^m), w\rangle = \langle (\sigma^m)^{-1}y, Dg_t^m(w)w\rangle$$

Thus

$$\|\langle h^m R_N(\sigma^m), w\rangle\|_{L^1(\gamma)} \le (N+1)\,\|y\|\,\sup_m E\,\|\eta_t\|_{L^1(\gamma)}$$

for each m where η $(=\eta^m)$ is the path $Dg^m(w)w$. This satisfies the equation

$$\eta_{kt/m} = \sum_{i=0}^{k-1}\{A(v_{it/m})\Delta_i w + DA(v_{it/m})(\eta_{it/m}, \Delta_i w)$$
$$+ t/m DB(v_{it/m})\eta_{it/m}\}$$

for $k = 1, \ldots, m$.

Since A, DA and DB are bounded, Lemma 4.5 gives

$$E\,\|\eta_{kt/m}\|^2 \le d_{19} \quad \text{for} \quad k = 1, 2, \ldots, m$$

and it follows that $E\,\|\eta_t\| \le d_{19}$ for all m.

It remains to show that

$$\sup_m \text{Trace}\,\|D[h^m R_N(\sigma^m)]\|_{L^1(\gamma)} < \infty.$$

For each m, we evaluate the trace on the basis B_m defined in (4.10). This gives

$$\text{Trace}\,D[h^m R_N(\sigma^m)] = \sum_{r=1,k=0}^{n,m-1} \langle D[h^m R_N(\sigma^m)]f^{rk}, f^{rk}\rangle_{L_0^{2,1}}$$
$$= \sum_{r=1,k=0}^{n,m-1} \{R_N(\sigma^m)[\langle(\sigma^m)^{-1}y, D^2 g_t^m(w)(f^{rk}, f^{rk})\rangle$$
$$- \langle(\sigma^m)^{-1} D\sigma^m(w) f^{rk}(\sigma^m)^{-1}y, Dg_t^m(w)f^{rk}\rangle]$$
$$+ DR_N(\sigma^m) D\sigma^m(w) f^{rk}\langle(\sigma^m)^{-1}y, Dg_t^m(w)f^{rk}\rangle\}.$$

Since R_N, $R_N(\sigma^m)(\sigma^m)^{-1}$ and $DR_N(\sigma^m)(\sigma^m)^{-1}$ are all bounded, it suffices to show that

$$\sup_m E\left\|\sum_{k=0}^{m-1} D^2 g_t^m(w)(f^{rk}, f^{rk})\right\| < \infty \tag{4.20}$$

and

$$\sup E \sum_{k=0}^{m-1} \|D\sigma^m(w)f^{rk}\| \, \|Dg_t^m(w)f^{rk}\| < \infty \tag{4.21}$$

for each $r = 1, 2, \ldots, n$.

As before let η^{rk} denote $\sqrt{(m/t)}Dg^m(w)f^{rk}$. Lemma 4.5 applied to equation (4.14) yields

$$E\,\|\eta_{jt/m}^{rk}\|^4 \le d_{20} \quad \text{for each} \quad j = 1, 2, \ldots, m \tag{4.22}$$

where d_{20} is independent of k and m. Let ρ^{rk} denote the path $m \cdot D^2g^m(w)(f^{rk}, f^{rk})$. This satisfies the equation

$$\begin{aligned}\rho_{jt/m}^{rk} &= tDA(v_{kt/m})(\eta_{kt/m}^{rk}, x_r)\\ &\quad + \sum_{i=k+1}^{j-1} \{tD^2A(v_{it/m})(\eta_{it/m}^{rk}, \eta_{it/m}^{rk}, \Delta_i w)\\ &\quad + DA(v_{it/m})(\rho_{it/m}^{rk}, \Delta_i w)\}\\ &\quad + \frac{t}{m}\sum_{i=k+1}^{j-1} \{tD^2B(v_{it/m})(\eta_{it/m}^{rk}, \eta_{it/m}^{rk}) + DB(v_{it/m})\rho_{it/m}^{rk}\}\end{aligned}$$

Applying Lemma 4.5 again in conjunction with (4.22) gives

$$E\,\|\rho_{jt/m}^{rk}\|^2 \le d_{21} \quad \text{for each} \quad j = 1, 2, \ldots, m$$

where d_{21} is independent of k and m. This implies (4.20).

We have shown above that the sequence $\{\sqrt{m}\,Dg_t^m(w)f^{rk} : 0 \le k \le m-1;\ m = 1, 2, \ldots\}$ is bounded in L^2. Hence (4.21) will follow if we show that the same is true for $\sqrt{m}\,D\sigma^m(w)f^{rk}$. Observe that

$$\begin{aligned}&[D\sigma^m(w)f^{rk}]_{ij} =\\ &\sum_{r,s=1;\ell,k=0}^{n;m-1} \{\langle z_i, D^2g_t^m(w)(f^{s\ell}, f^{rk})\rangle \langle z_j, Dg_t^m(w)f^{s\ell}\rangle\\ &\quad + \langle z_i, Dg_t^m(w)f^{s\ell}\rangle \langle z_j, D^2g_t^m(w)(f^{rk}, f^{s\ell})\rangle\}.\end{aligned}$$

Using exactly the same argument as before, one can show that the terms $E\,\|mD^2g_t^m(w)(f^{s\ell}, f^{rk})\|^4$ are bounded in ℓ, k and m. We have shown in (4.22) that $E\,\|\sqrt{m}\,Dg_t^m(w)f^{rk}\|^4$ are bounded with respect to k and m. It follows from Lemma 4.4(ii) together with the Cauchy–Schwartz inequality that the sequence $\{E\,\|\sqrt{m}\,[D\sigma^m(w)f^{rk}]\|^2\ k = 0, \ldots, m-1;\ m = 1, 2, \ldots\}$ is bounded as required. Hence (4.21) holds.

With this we have verified (4.19) and the proof of the theorem is complete. ∎

Theorem 4.10 *Suppose that A and B are bounded, have bounded derivatives of all orders, σ is invertible with probability* 1 *and $\sigma^{-1} \in L^p$ for all $p \ge 1$.*

Then the density of ξ_t is a C^∞ function.

Remark Cf. Theorems 2.5 and 3.4.

Proof Let ϕ, b, and $y_1, y_2, \ldots, y_b$ be as in Lemma 1.14. We will derive the estimate in (1.8) by iterating the procedure used in the proof of Theorem 4.9.

For each $j = 1, 2, \ldots, b$ define ϕ_j on $\mathbf{R}^d$ by

$$\phi_j(z) = D^j\phi(z)(y_1, y_2, \ldots, y_j)$$

and h_j^m on C_0 by

$$\begin{cases} h_j^m = Dg_t^m(w)^*[(\sigma^m)^{-1}y_j] & \text{if } \sigma^m \in GL(d) \\ 0 & \text{otherwise.} \end{cases}$$

Proceeding as in the proof of Theorem 4.9 one obtains

$$\int_{\mathbf{R}^d} D^b(z)(y_1, y_2, \ldots, y_b)\, d\nu(z) = \int_{C_0} \phi_b(g_t(w))\, d\gamma(w)$$

$$= \lim_{N\to\infty} \int_{C_0} \phi_b(g_t(w))R_N(\sigma)\, d\gamma(w)$$

$$= \lim_{N\to\infty} \lim_{m\to\infty} \int_{C_0} \phi_b(g_t^m(w))R_N(\sigma^m)\, d\gamma(w) \tag{4.23}$$

$$= \lim_{N\to\infty} \lim_{m\to\infty} \int_{C_0} D\phi_{b-1}(g_t^m(w))y_bR_N(\sigma^m)\, d\gamma(w)$$

$$= \lim_{N\to\infty} \lim_{m\to\infty} \int_{C_0} D(\phi_{b-1}\circ g_t^m)(w)h_b^mR_N(\sigma^m)\, d\gamma(w)$$

$$= \lim_{N\to\infty} \lim_{m\to\infty} \int_{C_0} \phi_{b-1}(g_t^m(w))\, \mathrm{Div}[h_b^mR_N(\sigma^m)]\, d\gamma(w). \tag{4.24}$$

Iterating between (4.23) and (4.24) eventually leads to the equality

$$\int_{\mathbf{R}^d} D^b\phi(z)(y_1, y_2, \ldots, y_b)\, d\nu(z)$$

$$= \lim_{N\to\infty} \lim_{m\to\infty} \int_{C_0} \phi(g_t^m(w))\, \mathrm{Div}[h_1^m\, \mathrm{Div}[h_2^m \ldots \mathrm{Div}[h_b^mR_N(\sigma^m)] \ldots]\, d\gamma(w).$$

Hence (1.8) will follow from the condition

$$\sup_N \sup_m E\, \|\mathrm{Div}[h_1^m\, \mathrm{Div}[h_2^m \ldots \mathrm{Div}[h_b^mR_N(\sigma^m)] \ldots]\| < \infty.$$

We will now show that this holds.

The expression inside the expectation above is dominated by a polynomial in the following terms

$$A_{N,m} \equiv \|D^j R_N(\sigma^m)\| \, \|(\sigma^m)^{-1}\| \quad \text{if} \quad \sigma^m \in GL(d)$$
$$\qquad 0 \qquad\qquad \text{otherwise}$$

where $0 \leq j \leq b$.

$$C_m \equiv \sum_{\ell_1=0,\,\ell_2=0,\ldots,\,\ell_{(n_1+n_2+\ldots n_t)/2}=0}^{m-1,m-1,\ldots,m-1} \{\|D^{n_1}G_1(w)(f^{i'_1 i_1}, \ldots, f^{i'_{n_1} i_{n_1}})\|$$
$$\times \|D^{n_2}G_2(w)(f^{j'_1 j_1}, \ldots, f^{j'_{n_2} j_{n_2}}\| \ldots \|D^{n_t}G_t(w)(f^{k'_1 k_1}, \ldots, f^{k'_{n_t} k_{n_t}})\|\}$$

here each function G_i $(1 \leq i \leq t)$ is one of $\{g_t^m, Dg_t^m(\cdot)w, \sigma^m\}$; each of the dashed indices is between 1 and n, and the set of undashed indices $\{i_1, \ldots, i_{n_1}, j_1, \ldots, j_{n_2}, \ldots, k_1, \ldots, k_{n_t}\}$ have been paired and the uth pair denoted by ℓ_u. For each term C_m of this form, $n_1 + n_2 + \ldots + n_t \leq 2b$; hence the number of such terms depends only on b. It is therefore sufficient to prove that for every p (a positive integral power of 2), both $\sup_N \lim_m E[A^p_{N,m}]$ and $\sup_m E[C^p_m]$ are finite.

Now for each fixed N, $A_{N,m}$ is bounded, $D^j R_N$ is continuous and $\sigma^m \to \sigma$ in measure as m tends to infinity; so by the dominated convergence theorem and (4.16)

$$E[A^p_{N,m}] \to E[\|D^j R_N(\sigma)\| \, \|\sigma^{-1}\|]^p \leq C^p_j E[\|\sigma^{-1}\|^{p(q+1)}]$$

as $m \to \infty$. This last term is finite under the hypothesis of the theorem.

It only remains to show that $E[C^p_m]$ is finite for every power p of 2. By Lemma 4.4(ii) and the Cauchy–Schwartz inequality this will be implied by the following result: For every q (a power of 2), integer s and $1 \leq i'_1, i'_2, \ldots, i'_s \leq n$

$$\sup_{m;0 \leq i_1, i_2, \ldots, i_s \leq m-1} E \, \|m^{s/2} D^s G(w)(f^{i'_1 i_1}, \ldots, f^{i'_s i_s})\|^q < \infty$$

where G is any one of the functions g_t^m, $Dg_t^m(\cdot)w$, σ^m. We will prove this in the first case by an inductive argument, noting that a similar argument also works for the other two functions. Make the following inductive assumption: for every q and $r \leq s-1$, there exists a constant K depending only on s and q such that

$$E \, \|m^{r/2} D^r g^m(w)(f^{i'_1 i_1}, \ldots, f^{i'_r i_r})_{jt/m}\|^q \leq \mathrm{K} \tag{4.25}$$

for every m; $0 \leq i_1, i_2, \ldots, i_r \leq m-1$ and $j = 1, 2, \ldots, m$.

Successive differentiation with respect to w in equation (4.2) shows that the path $m^{s/2} D^s g^m(w)(f^{i'_1 i_1}, \ldots, f^{i'_s i_s})$, which we denote below by η, satisfies an equation of the form

$$\begin{aligned}\eta_{jt/m} = {} & \{D^{\alpha}A(v_{bt/m})(\eta_{bt/m}^{(1)}, \eta_{bt/m}^{(2)}, \ldots, \eta_{bt/m}^{(\alpha-1)}, x_d) + \ldots\} \\ & + \sum_{i=k+1}^{j-1} \{D^{\beta}A(v_{it/m})(\eta_{it/m}^{(\alpha)}, \eta_{it/m}^{(\alpha+1)}, \ldots, \eta_{it/m}^{(\alpha+\beta-2)}, \Delta_i w) + \ldots\} \\ & + \frac{t}{m} \sum_{i=k+1}^{j-1} \{D^{\gamma}B(v_{it/m})(\eta_{it/m}^{(\alpha+\beta-1)}, \eta_{it/m}^{(\alpha+\beta)}, \ldots, \eta_{it/m}^{(\alpha+\beta+\gamma-2)}) \\ & + \ldots\} \\ & + \sum_{i=k+1}^{j-1} DA(v_{it/m})(\eta_{it/m}, \Delta_i w) + \frac{t}{m} \sum_{i=k+1}^{j-1} DB(v_{it/m})\eta_{it/m} \end{aligned} \tag{4.26}$$

for $j = 1, 2, \ldots, m$; where each $\eta^{(\delta)}$, $1 \le \delta \le \alpha + \beta + \gamma - 2$, is a path of the form

$$m^{r/2} D^r g^m(w)(f^{i_1' i_1}, \ldots, f^{i_r' i_r}); \qquad r \le s - 1,$$

and the number of terms in each of the above brackets depends only upon s. All terms in the same bracket have the same form and we have written down a typical example. Then by applying Lemma 4.5 and the Cauchy–Schwartz inequality to (4.26) and making use of the inductive assumption (4.25), we see that (4.25) also holds for $r = s$.

Finally note that the application of Lemma 4.5 to equation (4.14) gives (4.25) in the case $r = 1$. This completes the proof of Theorem 4.10. ■

5

A discussion of the different forms of the theory

5.1 An outline of Malliavin's original paper

The development of the Malliavin calculus was motivated by the following result from finite dimensional differential analysis.

Theorem 5.1 *Suppose that $T: E \to F$ is a C^∞ map between Euclidean spaces. Let γ be a measure on E and denote by ν the induced measure $T(\gamma)$ on F. Let λ denote Lebesgue measure on F and define $N \equiv \{x \in E: DT(x)$ is non-surjective$\}$. Then*
(i) *$\nu \ll \lambda$ implies $\gamma(N) = 0$.*
(ii) *Suppose $\gamma(N) = 0$ and for every vector $y \in F$, there exists a random variable h_y on (E, γ) such that $DT(x)h_y(x) \equiv y$ which lies in the domain of the operator D^* (where D denotes differentiation of real-valued C^1 functions on E and D^* the formal adjoint of D with respect to $L^2(\gamma)$). Then $\nu \ll \lambda$.*

Part (i) is a direct consequence of Sard's theorem which asserts that $\lambda(T(N)) = 0$. However, since $\nu(T(N)) \geq \gamma(N)$, the condition $\gamma(N) > 0$ would imply $\nu(T(N)) > 0$ and hence that $\nu \not\ll \lambda$.

Part (ii) can be obtained from the argument used to prove absolute continuity of $F(\gamma_m)$ in section 4.1.

Thus in the case where E is finite dimensional the regularity of the induced measure ν is related to a property (i.e. surjectivity) of the *Frechet derivative* of T. In attempting to extend this result to the situation where ν is the measure induced by the solution ξ_t of a stochastic differential equation, one is faced with the problem of the non-differentiability of the map $g_t: w \to \zeta_t$.

Malliavin introduces a level of differentiability into the problem by the device of embedding the Wiener measure γ into the distribution of a stationary stochastic process $\{x(\tau)=x_t(\tau):\tau\in\mathbf{R}\}$. This process is constructed so that $g_t(x(\tau))$ is differentiable with respect to the parameter τ. The estimates required by Lemma 1.14 are then obtained by computations which involve differentiation with respect to τ. That this procedure forms the basis of the work of both Stroock and Bismut is the justification for regarding both methods as forms of the same theory.

Malliavin implements his program in reference 19 in the following way.† He first constructs a C_0-valued Ornstein–Uhlenbeck process $x=\{x\,.\,(\tau):\tau\in\mathbf{R}\}$ by defining

$$x_t(\tau)=\int_{-\infty}^{\tau}\int_0^t e^{-1/2(s-\tau)}\,dv(s,r);\qquad t\in[0,1],\qquad \tau\in\mathbf{R}$$

where v is white noise defined on the strip $\{(t,\tau):0\leq t\leq 1,\ \tau\in\mathbf{R}\}$. x is a stationary Markov process with the Wiener measure γ as distribution at each time τ. It is also invariant under time reversal in the sense that for any τ_0 the distributions of $\{x(\tau):\tau\in\mathbf{R}\}$ and $\{x(\tau_0-\tau):\tau\in\mathbf{R}\}$ coincide.

A random variable u in $L^2(\gamma)$ is considered to be regular in Malliavin's sense if there exists k in $L^2(\gamma)$ such that

$$u(x(\tau))-\int_0^{\tau}k(x(s))\,ds,\qquad \tau\geq 0$$

is a martingale with respect to the parameter τ. If k exists, then it is necessarily unique up to stochastic equivalence. An operator A is defined on the set of functionals u in $L^2(\gamma)$ for which a corresponding k exists by $A(u)=k$. An equivalent description of A is as the closed extension in $L^2(\gamma)$ of the generator of the Markov process $\{x(\tau):\tau\in\mathbf{R}\}$ (see Stroock[27] for details of this). Thus for u in the domain of A

$$Au(x(0))=\lim_{\tau\downarrow 0}E\left[\frac{u(x(\tau))-u(x(0))}{\tau}\Big/x(0)\right]$$

This is the differentiability in τ which we refer to above. The process $\{x(\tau):\tau\geq 0\}$ is equivalent to the infinite dimensional Ornstein–Uhlenbeck process $\{\eta_\tau:\tau\geq 0\}$ in (2.12) and the operator A is precisely the symmetric diffusion operator $\mathscr{L}$ introduced in section 2.2.

Having defined A, and then an associated bilinear form $(u,v)\to\nabla u\cdot\nabla v$ ($\langle\,,\rangle$ of section 2.2), Malliavin shows that a random variable $f=(f_1,\ldots,f_d)$ satisfying certain hypotheses induces an absolutely continuous distribution on $\mathbf{R}^d$. The essential conditions required on f are that each of its components is in the domain of A (i.e. differentiable in

† We assume here that we are dealing with real-valued Brownian motion in order to simplify notation.

Malliavin's sense) and the matrix $(\nabla f_i \cdot \nabla f_j)_{i,j=1}^d$ is non-degenerate with probability 1. The computations that lead to this absolute continuity result are similar to those in the proof of Theorem 2.5 and make use of the fact that A is a symmetric operator on $L^2(\gamma)$. The symmetry of A is a consequence of the time-reversal invariance of the Ornstein–Uhlenbeck process x.

The argument is completed by showing that the solution map $g_t = (g_t^{(1)}, \ldots, g_t^{(d)})$ of a stochastic differential equation with smooth coefficients has all components $g_t^{(i)}$ in the domain of A. This is done by employing a mollification technique. Malliavin introduces a sequence of operators $\{T_\varepsilon : \varepsilon \geq 0\}$ which take continuous paths on $[0, 1]$ to smooth paths, defined so that $T_\varepsilon \to I$ as $\varepsilon \downarrow 0$. For each $\varepsilon > 0$ the action of A on the regularized maps $g_{t,\varepsilon}^{(i)} \equiv g_t^{(i)} \circ T_\varepsilon$ is computed. Then a powerful limit theorem (also contained in Malliavin[19]) is used to show that $g_{t,\varepsilon}^{(i)} \to g_t^{(i)}$, and $A(g_{t,\varepsilon}^{(i)})$ converges to a limit $f_t^{(i)}$ in $L^2(\gamma)$ as $\varepsilon \downarrow 0$. Since A is a closed operator, it follows that $g_t^{(i)}$ is in the domain of A and $Ag_t^{(i)} = f_t^{(i)}$. In this way one derives a stochastic differential equation for each process $f^{(i)}$ and the covariance matrix $\sigma_t = (\nabla g_t^{(i)} \cdot \nabla g_t^{(j)})_{i,j=1}^d$ can then be obtained by Itô's lemma. Hence σ also satisfies a stochastic differential equation. It can thus be shown that the operations A and $(u, v) \to \nabla u \cdot \nabla v$ may be applied arbitrarily often to the maps $g_t^{(i)}$ and $\sigma_{ij,t}$; $1 \leq i, j \leq d$. This enables the analysis to be used to study smoothness of the density of the induced measure ν as the proof of Theorem 2.5 illustrates.

Malliavin also outlined in reference 19 a method whereby Hörmander's conditions on the coefficients of the stochastic differential equation could be shown to imply the required non-degeneracy of the matrix σ. This was subsequently taken up and developed by other workers in the field.

5.2 A condition for equivalence of the approaches of Stroock and Bismut

Let (C_0, γ) denote the n-dimensional Wiener space and suppose that f is an L^2 map from C_0 to $\mathbf{R}^d$. Note that for any (bounded) adapted process $h: C_0 \to L_0^{2,1}$ and real number λ, $f(w + \lambda h(w))$ is a well defined random variable by the Girsanov theorem. Suppose that for every such process h the following limit exists in L^2

$$Df(w)h(w) = \lim_{\lambda \to 0} \frac{f(w + \lambda h(w)) - f(w)}{\lambda}.$$

Assume now that f also lies in the domain of the operator $\mathcal{L}$ defined in chapter 2. We will give a further condition on f under which the methods of Stroock and Bismut are equally applicable to the study of the regularity of the measure $f(\gamma)$. Recall that the hypotheses required in order to obtain absolute continuity of the induced measure by each of these approaches are, respectively:

(i) The matrix $\sigma \equiv (\langle f_i, f_j \rangle)_{i,j=1}^d$ is non-degenerate a.s.

(ii) There exists an adapted $n \times d$ matrix-valued process $\{H_s = [H_1(s), \ldots, H_d(s)] : s \in [0, 1]\}$ such that the $d \times d$ matrix

$$\tau \equiv \left[Df(w) \int_0^{\cdot} H_1(s)\,\mathrm{d}s, \ldots, Df(w) \int_0^{\cdot} H_d(s)\,\mathrm{d}s \right]$$

is non-degenerate a.s.

It can be shown that the former condition is actually necessary for the absolute continuity of $f(\gamma)$ (see Kusuoka[17]), which implies that (i) will hold whenever (ii) does. It is not clear to me whether or not the converse is true in general. However, (i) certainly implies (ii) if

(iii) The process $Df(w)^*$ is of the form

$$\left[\int_0^{\cdot} J_s\,\mathrm{d}s \right] Z$$

where J is an adapted $n \times d$ matrix-valued process in $L^2[0, 1]$ and Z is a $d \times d$ matrix, non-degenerate with probability 1.

Indeed if (iii) holds and one chooses H in (ii) to be the matrix J, then† $\tau = Df(w)Df(w)^* Z^{-1} = \sigma Z^{-1}$, so the non-degeneracy of τ follows from that of σ. The following result shows that this is always so if f is a map defined by a stochastic differential equation.

Theorem 5.2 *Suppose that $A : \mathbf{R}^d \to \mathbf{M}^{n,d}$ and $B : \mathbf{R}^d \to \mathbf{R}^d$ are bounded and C^2 with bounded first and second derivatives. Let ξ be the solution of the Stratonovich stochastic differential equation*

$$\xi_s = x + \int_0^s A(\xi_u)\,\mathrm{d} \circ w_u + \int_0^s B(\xi_u)\,\mathrm{d}u, \qquad s \in [0, 1]$$

and denote by f the map $w \to \xi_t$ for some $0 < t < 1$. Then f satisfies condition (iii).

Proof Define a map $\tilde{g} : L_0^{2,1} \to L_0^{2,1}$ by $\tilde{g}(y) = z$ and let $\tilde{g}_t(y) = z_t$,

† Note that $\langle , \rangle$ is an extension of the bilinear form $(F, G) \to DF(\cdot)DG(\cdot)^*$ defined for real-valued C^1 functions F and G on C_0.

where z is the solution of the integral equation

$$z_s = x + \int_0^s \{A(z_u)y'_u + B(z_u)\}\, du, \qquad u \in [0, 1] \tag{5.1}$$

Then $\tilde{g}$ is a Frechet differentiable map on $L_0^{2,1}$ and for any y and r in $L_0^{2,1}$ the path $\eta = D\tilde{g}(y)r$ satisfies the equation

$$\eta_s = \int_0^s \{DA(z_u)(\eta_u, y'_u) + A(z_u)r'_u + DB(z_u)\eta_u\}\, du, \qquad s \in [0, 1] \tag{5.2}$$

Furthermore if $\{P_m : m = 1, 2, \ldots\}$ denotes the sequence of piecewise linear projections in section 4.1, then $\tilde{g}_t \circ P_m$ converges to f in $L^2(\gamma)$ (see, for example, Elworthy).[8] We will compute $Df(w)^*$ as the limit of the sequence $Dg_t(P_m w)^*$.

Using (5.2) it is easy to show that for any vectors $h \in \mathbf{R}^d$ and $y \in L_0^{2,1}$, the path $D\tilde{g}_t(y)^*h$ is given by

$$D\tilde{g}_t(y)^*h = \int_0^{\cdot \vee t} A(z_u)^* \rho_u\, du \tag{5.3}$$

where $z = \tilde{g}(y)$, and ρ satisfies

$$\rho_s = h + \int_s^t [DA(z_u)(\cdot, y'_u)^* + DB(z_u)^*]\rho_u\, du, \qquad 0 < s \leq t$$

$$\rho_s = 0 \text{ if } s > t.$$

Let Z be the $d \times d$ matrix valued process defined by the equation

$$Z(s) = I - \int_0^s [DA(z_u)(\cdot, y'_u)^* + DB(z_u)^*]Z(u)\, du, \qquad 0 \leq s \leq 1 \tag{5.4}$$

Then it can be shown by writing down an explicit integral equation for the process $Z(\cdot)^{-1}$ that for each $0 \leq s \leq 1$ the matrix $Z(s)$ is invertible and ρ in (5.3) is given by

$$\begin{cases} \rho_s = Z(s)Z(t)^{-1}h, & 0 \leq s \leq t. \\ 0 & \text{otherwise} \end{cases}$$

Thus one obtains

$$D\tilde{g}_t(y)^* = \left[\int_0^{\cdot \vee t} A(z_u)^* Z(u)\, du\right] Z(t)^{-1}. \tag{5.5}$$

Now replace y by $P_m w$ in equations (5.4) and (5.5) and denote the corresponding matrix process Z by Z_m. As $m \to \infty$, Z_m converges to the process Z where Z is the solution of the Stratonovich equation

$$Z(s) = I - \int_0^s DA(\xi_u)(\cdot, \mathrm{d} \circ w_u)^* Z(u)$$
$$- \int_0^s DB(\xi_u)^* Z(u)\,\mathrm{d}u, \qquad 0 \le s \le 1$$

We will still have $Z(s) \in GL(d)$ a.s. for each s (the equation for $Z(\cdot)^{-1}$ can be obtained from Itô's lemma). Hence $Dg_t(p_m w)^*$ converges in L^2 to the process

$$Df(w)^* = \left[\int_0^{\cdot \vee t} A(\xi_u)^* Z(u)\,\mathrm{d}u\right] Z(t)^{-1}.$$

Thus (iii) holds. ■

5.3 Transformation theorems and integration by parts operators

Let E denote a Banach space such that E and E^* are both separable, equipped with a (not necessarily Gaussian) Borel measure γ. By a transformation theorem I mean a result which asserts that if T is a map from E to itself with certain properties, then the measure $T(\gamma)$ is absolutely continuous with respect to γ. This is a natural analogue of the type of regularity result with which we have been concerned in the previous chapters, in a situation where T ranges in an infinite dimensional vector space. One example of a transformation theorem is Ramer's result[25]. Another is the Girsanov theorem. In this section I would like to outline a scheme whereby a transformation theorem of the type described above might be obtained from a suitable version of the operator D^*, the formal adjoint with respect to $L^2(\gamma)$ of the differentiation operator. Let $\mathscr{D}$ denote a fixed subset of the class of functions on E.

Definition 5.3 A linear operator $\mathscr{A}$ from $\mathscr{D}$ to $L^2(\gamma)$ is an *integration by parts operator* (IPO) for γ if the relation

$$\int_E D\phi(x)h(x)\,\mathrm{d}\gamma(x) = \int_E \phi(x)(\mathscr{A}h)(x)\,\mathrm{d}\gamma(x) \tag{5.6}$$

holds for all C^1 functions ϕ from E to $\mathbf{R}$ and all $h \in \mathscr{D}$ for which either side in (5.6) exists.

For example, suppose that γ is a Gaussian measure on E arising from an abstract Wiener space (i, H, E). Take as $\mathscr{D}$ the set $\mathscr{D}_1$ of all C^1 functions from E to E^* (where E^* is considered as a subspace of E under the inclusions $E^* \xrightarrow{i^*} H^* \cong H \xrightarrow{i} E$). For $h \in \mathscr{D}_1$ define

$$(\mathscr{A}_1 h)(x) = h(x)x - \mathrm{Trace}_H\, Dh(x).$$

Then by Theorem 1.4 $\mathscr{A}_1$ is an IPO.† However, if one is dealing with the classical Wiener space, then there exists an IPO of an entirely different kind.

Let $\mathscr{D}_2 = \{h : C_0 \to L_0^{2,1}$ such that h is bounded and non-anticipating$\}$.

Let $\phi : C_0 \to \mathbf{R}$ be any C^1 function such that ϕ and $D\phi$ are both bounded. For $h \in \mathscr{D}_2$ the Girsanov theorem gives

$$\int_{C_0} \phi(x + \lambda h(x))\, d\gamma(x)$$
$$= \int_{C_0} \phi(x) \exp\left\{\lambda \int_0^1 h_s'\, dx_s - \lambda^2/2 \int_0^1 h_s'^2\, ds\right\} d\gamma(x).$$

Differentiating with respect to λ, then setting $\lambda = 0$, gives

$$\int_{C_0} D\phi(x)h(x)\, d\gamma(x) = \int_{C_0} \phi(x)(\mathscr{A}_2 h)(x)\, d\gamma(x) \tag{5.7}$$

where $(\mathscr{A}_2 h)(x)$ is the Itô integral $\int_0^1 h_s'\, dx_s$. This relation holds for more general ϕ by an approximation argument. Thus $\mathscr{A}_2$ satisfies (5.6).‡ Note that definition of $\mathscr{A}_2$ does not involve differentiation of h with respect to x but uses the function space structure of C_0 in an essential way. In the case where h is both smooth and non-anticipating, then the trace term in $\mathscr{A}_1 h$ vanishes, and since $h(x)x$ is an abstract version of the Itô integral this implies $\mathscr{A}_1 h = \mathscr{A}_2 h$.

Suppose now that γ has an IPO $\mathscr{A}$ with domain $\mathscr{D}$. Then $\mathscr{A}$ satisfies the following:

Lemma 5.4 *Let $h \in \mathscr{D} \cap L^2(\gamma)$, $\psi : E \to \mathbf{R} \in L^2 \cap C^1$ and suppose $\psi h \in \mathscr{D}$. Then*

$$\mathscr{A}(\psi h)(x) = \psi(x)\mathscr{A}h(x) - D\psi(x)h(x) \quad a.s.\ (\gamma) \tag{5.8}$$

† Note that Stroock's operator $\mathscr{L}$ is an extension of $\mathscr{A}_1 \circ D$.

‡ This has also been observed by Gaveau and Trauber in reference 10.

Proof Let $\phi: E \to \mathbf{R}$ be a bounded C^1 function with bounded derivative. Then

$$\int_E \phi(x)\psi(x)\mathscr{A}h(x)\,\mathrm{d}\gamma(x)$$

$$= \int_E \{\psi(x)D\phi(x)h(x) + \phi(x)D\psi(x)h(x)\}\,\mathrm{d}\gamma(x)$$

$$= \int_E \{D\phi(x)[\psi(x)h(x)] + \phi(x)D\psi(x)h(x)\}\,\mathrm{d}\gamma(x)$$

$$= \int_E \phi(x)\{\mathscr{A}[\psi h](x) + D\psi(x)h(x)\}\,\mathrm{d}\gamma(x).$$

Thus (5.8) follows. ■

Now let $T: E \to E$ be a map of the form $T = I + K$, where I is the identity and $K \in \mathscr{D}$. Define a homotopy between I and T by

$$T_t = I + tK, \qquad t \in [0, 1].$$

Suppose that T_t is invertible for each $t \in [0, 1]$. Observe that $T_t(\gamma) \ll \gamma$ for each t if and only if there exists a family $\{X_t : t \in [0, 1]\}$ of $L^1(\gamma)$ random variables (i.e. the Radon–Nikodym derivatives $\mathrm{d}T_t(\gamma)/\mathrm{d}\gamma$) such that

(i) $X_0 \equiv 1$.

(ii) For each test function ϕ on E

$$\int_E \phi(x)\,\mathrm{d}\gamma(x) = \int_E \phi \circ T_t^{-1}(x)X_t(x)\,\mathrm{d}\gamma(x) \tag{5.9}$$

for all $t \in [0, 1]$.

We will derive formulae ((5.12) below) for $\{X_t : t \in [0, 1]\}$ in terms of $\mathscr{A}$, assuming that these are sufficiently regular and that certain formal manipulations are valid.

Assume that (i) and (ii) above hold and that the map $(t, x) \to X_t(x)$ is differentiable in t and x. Let ϕ be as in (ii), and denote the right-hand side in (5.9) by $f(t)$. Then $f'(t) \equiv 0$. Differentiation with respect to t under the integral sign gives

$$0 \equiv \int_E \Big\{D\phi(T_t^{-1}(x))\frac{\mathrm{d}}{\mathrm{d}t}T_t^{-1}(x) \cdot X_t(x)$$
$$+ \phi \circ T_t^{-1}(x)\frac{\mathrm{d}}{\mathrm{d}t}X_t(x)\Big\}\,\mathrm{d}\gamma(x).$$

The first term in the integrand may be simplified by using the following relations

$$D\phi(T_t^{-1}(x))\frac{\mathrm{d}}{\mathrm{d}t}T_t^{-1}(x)$$

$$= D(\phi \circ T_t^{-1})(x)[DT_t^{-1}(x)]^{-1}\frac{\mathrm{d}}{\mathrm{d}t}T_t^{-1}(x)$$

$$= D(\phi \circ T_t^{-1})(x)DT_t(T_t^{-1}(x))\frac{\mathrm{d}}{\mathrm{d}t}T_t^{-1}(x)$$

$$= -D(\phi \circ T_t^{-1})(x)\frac{\mathrm{d}}{\mathrm{d}t}(T_t)(T_t^{-1}(x))$$

$$= -D(\phi \circ T_t^{-1})(x)K \circ T_t^{-1}(x).$$

Assume that for each $t \in [0, 1]$, $K \circ T_t^{-1} \cdot X_t \in \mathscr{D}$. Then by (5.6) we will have

$$0 \equiv \int_E \phi \circ T_t^{-1}(x)\left\{\frac{\mathrm{d}}{\mathrm{d}t}X_t(x) - \mathscr{A}[K \circ T_t^{-1} \cdot X_t](x)\right\}\mathrm{d}\gamma(x).$$

Since this holds for every test function ϕ on E, it follows that

$$\frac{\mathrm{d}}{\mathrm{d}t}X_t(x) - \mathscr{A}[K \circ T_t^{-1} \cdot X_t](x) \equiv 0 \quad \text{a.s.} \quad (\gamma). \tag{5.10}$$

Suppose that for each $t \in [0, 1]$, $K \circ T_t^{-1}$ and X_t satisfy the conditions on h and ψ in Lemma 5.4. Then applying (5.8) to equation (5.10) gives

$$\frac{\mathrm{d}}{\mathrm{d}t}X_t(x) - X_t(x)\mathscr{A}[K \circ T_t^{-1}](x) + DX_t(x)K \circ T_t^{-1}(x) \equiv 0$$

Now write $X_t(x) = X(t, x)$, $X_1(t, x) = \mathrm{d}/\mathrm{d}t\, X_t(x)$, $X_2(t, x) = D_x X_t(x)$ and make the substitution $x = T_t(y)$. Then the previous equation reads

$$\begin{aligned} X_1(t, T_t(y)) - X(t, T_t(y))\mathscr{A}[K \circ T_t^{-1}](T_t(y)) \\ + X_2(t, T_t(y))K(y) \equiv 0 \end{aligned} \tag{5.11}$$

However, since $K = \mathrm{d}/\mathrm{d}t\, T_t$, equation (5.11) reduces to

$$\frac{\mathrm{d}}{\mathrm{d}t}X(t, T_t(y)) = X(t, T_t(y))\mathscr{A}[K \circ T_t^{-1}](T_t(y)).$$

Together with the condition $X(0, x) \equiv 1$, this implies

$$X(t, T_t(y)) = \exp\left[\int_0^t \mathscr{A}[K \circ T_s^{-1}](T_s(y))\,\mathrm{d}s\right].$$

We therefore arrive at the following expression for X

$$X(t, x) = \exp\left[\int_0^t \mathscr{A}[K \circ T_s^{-1}](T_s \circ T_t^{-1}(x))\, ds\right]. \tag{5.12}$$

In particular

$$\frac{d\nu}{d\gamma}(x) = \exp\left[\int_0^1 \mathscr{A}[K \circ T_s^{-1}](T_s \circ T^{-1}(x))\, ds\right] \tag{5.13}$$

where ν denotes the measure $T(\gamma)$.

Conversely if $T_t(\gamma) \ll \gamma$ for each $t \in [0, 1]$ and $dT_t(\gamma)/d\gamma = X_t$, then the argument used to produce (5.7) above shows that dX_t/dt evaluated at $t = 0$ is an IPO for γ.

Let $\mathscr{A}_1$ and $\mathscr{A}_2$ be as defined earlier. The formula above yields the density in Ramer's theorem in the case where γ is an abstract Gaussian measure and $\mathscr{A}$ in (5.13) is chosen to be $\mathscr{A}_1$. If γ is the classical Wiener measure on C_0 and $\mathscr{A}_2$ is used for $\mathscr{A}$ in (5.13), then the Girsanov density is obtained.

Suppose finally that an operator $\mathscr{A}$ with domain $\mathscr{D}$ is an IPO for γ, and $K: E \to E$ is a C^1 map lying in $\mathscr{D}$ for which the family of maps on E, $\{T_t = I + tK,\ t \in [0, 1]\}$ are invertible. Define $X(t, x)$ as in (5.12). Then in principle one could reverse the argument above, arrive at (5.9) and thereby prove a transformation theorem for ν. This was done in Bell[2] in the very specific case where K is constant. In this case the method yields a generalization of the Cameron–Martin theorem (cf. Theorem 7.10). More generally the regularity conditions required on $X(t, x)$ in order to justify the formal steps in the argument prove to be somewhat restrictive. However, this difficulty might be overcome by introducing a suitable extension of the derivative on E, in the spirit of the preceding chapters.

6

Non-degeneracy of the covariance matrix under Hörmander's condition

6.1 Hörmander's theorem

Throughout this chapter $A_0, A_1, \ldots, A_n$ will denote $n+1$ bounded C^∞ vector fields on $\mathbf{R}^d$ with bounded derivatives of all orders. Let x be a point in $\mathbf{R}^d$. Hörmander studied the second order Cauchy problem associated with $A_0, \ldots, A_n$ under the assumption that the vectors

$$\begin{aligned}&A_i(x), [A_j, A_k](x), [[A_j, A_k], A_l](x), \ldots \\ &i \in \{1, \ldots, n\}, \qquad j, k, l, \ldots \in \{0, \ldots, n\} \text{ span } \mathbf{R}^d\end{aligned} \tag{6.1}$$

Our goal in this chapter is to prove the following probabilistic form of Hörmander's theorem.

Theorem 6.1 *Suppose that ξ is the solution of the Stratonovich equation*

$$\xi_t = x + \int_0^t A_0(\xi_s)\,\mathrm{d}s + \sum_{i=1}^n \int_0^t A_i(\xi_s)\,\mathrm{d} \circ w_i(s), \qquad t \in [0, 1] \tag{6.2}$$

where $w = (w_1, \ldots, w_n)$ is Brownian motion in $\mathbf{R}^n$ and $A_0, \ldots, A_n$ satisfy (6.1).

Then for every $t \in (0, 1]$, the random variable ξ_t has an absolutely continuous distribution with a smooth (C^∞) density.

Before proving Theorem 6.1, I would like to present a heuristic argument which provides some insight into the relationship between the hypothesis and conclusion of the theorem. In order to simplify notation I will assume that we are dealing with only two vector fields A_1 and A_2 and the equation

$$\xi_t = x + \int_0^t A_1(\xi_s)\,\mathrm{d} \circ w_1(s) + \int_0^t A_2(\xi_s)\,\mathrm{d} \circ w_2(s).$$

Suppose that we wish to obtain the conclusion of Theorem 6.1. Note that initially ξ_t moves away from x in the plane P, say, spanned by the vectors $A_1(x)$ and $A_2(x)$. However, since it is clear that ξ_t will not induce an absolutely continuous distribution on $\mathbf{R}^d$ if it stays in a hyperspace of lower dimension than d, it follows (at least if $d \geq 3$) that ξ_t must be able to escape from P. Furthermore, since the theorem is to hold for every $t \in (0, 1]$, this must happen at arbitrarily small times t. Now one may effectively assume that in an arbitrarily small time interval $[0, t_4]$ the Brownian motion $w = (w_1, w_2)$ will describe a square path in $\mathbf{R}^2$ with sides of infinitesimal length which we denote by δ.

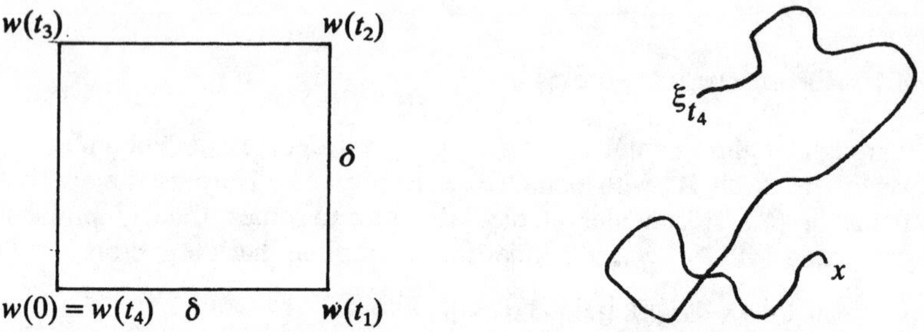

The solution will move from x to a new point ξ_{t_4} which can be computed as follows. For $s \leq t_1$, Taylor's formula gives

$$A_1(\xi_s) = A_1(x) + DA_1(x)A_1(x)w_1(s) + \dots .$$

Substituting this into the stochastic differential equation above, we obtain

$$\xi_{t_1} = x + A_1(x)\delta + DA_1(x)A_1(x)\delta^2/2 + o(\delta^2).$$

Now consider the path of the solution between times t_1 and t_2. Here there is no variation in w_1, and w_2 moves between 0 and δ. Work the same calculation with A_2 in place of A_1 and ξ_{t_1} in place of x, then use Taylor's formula again. The result is

$$\begin{aligned}\xi_{t_2} &= \xi_{t_1} + A_2(\xi_{t_1})\delta + DA_2(\xi_{t_1})A_2(\xi_{t_1})\delta^2/2 + o(\delta^2)\\ &= x + A_1(x)\delta + DA_1(x)A_1(x)\delta^2/2 + A_2(x)\delta\\ &\quad + DA_2(x)A_1(x)\delta^2 + DA_2(x)A_2(x)\delta^2/2 + o(\delta^2).\end{aligned}$$

Repeating this for the time interval $[t_2, t_3]$, we have

$$\begin{aligned}\xi_{t_3} &= \xi_{t_2} - A_1(\xi_{t_2})\delta + DA_1(\xi_{t_2})A_1(\xi_{t_2})\delta^2/2 + o(\delta^2)\\ &= x + A_2(x)\delta + [A_2, A_1](x)\delta^2\\ &\quad + DA_2(x)A_2(x)\delta^2/2 + o(\delta^2).\end{aligned}$$

Finally a similar calculation on the time interval $[t_3, t_4]$ gives

$$\begin{aligned}\xi_{t_4} &= \xi_{t_3} - A_2(\xi_{t_3})\delta + DA_2(\xi_{t_3})A_2(\xi_{t_3})\delta^2/2 + o(\delta^2) \\ &= x + [A_2, A_1](x)\delta^2 + o(\delta^2).\end{aligned}$$

We have thus shown that ξ_t is also driven away from x in the direction of the vector $[A_2, A_1](x)$. By iterating this argument one can actually show that in any infinitesimal neighbourhood of the time $t = 0$, the solution ξ proliferates in the directions determined by the vectors in (6.1). It is therefore natural to require that these vectors span $\mathbf{R}^d$, which is the hypothesis of Theorem 6.1.

Suppose from now on that condition (6.1) is satisfied, and let $t \in (0, 1]$ be a fixed time. It is easily shown that the covariance matrices σ $(= \sigma_t)$ associated with equation (6.2) which were introduced in chapters 2 and 4 are identical and may be written in the form

$$\sigma_t = Z_t^{-1}\left[\int_0^t Z_s A(\xi_s)A(\xi_s)^* Z_s^* \, ds\right][Z_t^{-1}]^*$$

where $A = (A_1, \ldots, A_n)$ and Z is defined by the equation

$$Z_s = I - \int_0^s Z_u DA_0(\xi_u)\, du - \sum_{i=1}^n \int_0^s Z_u DA_i(\xi_u)\, d \circ w_i(u), \qquad s \in [0, t].$$

Here I is the $d \times d$ identity matrix. The matrix η_t in chapter 3 is equal to $\sigma_t Z_t^*$. Let us now use σ to denote $\int_0^t Z_s A(\xi_s)A(\xi_s)^* Z_s^* \, ds$. By writing the integral equation for the process $\{Z_s^{-1}\}$ and then using Theorem 1.9(iv) to produce estimates on the L^p-norms of these matrices, one can show that Z_t^{-1} is in L^p for every p. The same is true for Z_t. Hence, in view of any one of Theorem 2.10, Theorem 3.4 or Theorem 4.10, the conclusion of Theorem 6.1 will follow from:

Theorem 6.2 *The matrix σ is invertible a.s. and $\sigma^{-1} \in L^p$ for all $P \in \mathbf{N}$.*

In this section we will prove the first part of this result, i.e. that (6.1) implies that σ is almost surely invertible. Note that by Theorem 4.9 this allows us to conclude that ξ_t has an absolutely continuous distribution.

Lemma 6.3 *Suppose that B is a C^2 vector field on $\mathbf{R}^d$ and τ is a stopping time such that*

$$Z(s)B(\xi_s) \equiv 0 \qquad \text{for } s \in [0, \tau] \tag{6.3}$$

Then for each $i = 0, \ldots, n$

$$Z(s)[A_i, B](\xi_s) \equiv 0 \qquad \text{on } [0, \tau].$$

Proof (6.3) implies

$$d(Z(s)B(\xi_s)) \equiv 0 \qquad \text{for } s \in [0, \tau].$$

Using Itô's lemma to evaluate the left-hand side gives

$$Z_s[B, A_0](\xi_s)\,ds + \sum_{i=1}^{n} Z_s[B, A_i](\xi_s)\,d \circ w_i(s) \equiv 0, \qquad s \in [0, \tau].$$

Writing the stochastic differentials in Itô form, we obtain

$$\left\{Z_s[B, A_0](\xi_s) + \frac{1}{2}\sum_{i,j=1}^{n} Z_s[[B, A_i], A_j](\xi_s)\right\} ds$$
$$+ \sum_{i=1}^{n} Z_s[B, A_i](\xi_s)\,dw_i(s) \equiv 0, \qquad s \in [0, \tau].$$

It follows that

$$Z_s[B, A_i](\xi_s) \equiv 0, \qquad s \in [0, \tau]; \qquad i = 1, \ldots, n$$

and

$$Z_s[B, A_0](\xi_s) + \frac{1}{2}\sum_{i,j=1}^{n} Z_s[[B, A_i], A_j](\xi_s) \equiv 0, \qquad s \in [0, \tau].$$

Iterating this argument on the first relation gives $Z_s[[B, A_i], A_j](\xi_s) \equiv 0$ on $[0, \tau]$ for all $i, j \in \{1, \ldots, n\}$. Using this in the second relation then gives

$$Z_s[B, A_0](\xi_s) \equiv 0 \quad \text{on } [0, \tau].$$

Hence the conclusion follows. ∎

Note that we could have replaced the hypothesis and conclusion of Lemma 6.3 by the statements $\langle Z(s)B(\xi_s), y\rangle \equiv 0$ for $s \in [0, \tau]$ and $\langle Z(s)[A_i, B](\xi_s), y\rangle \equiv 0$ on $[0, \tau]$ for any $y \in \mathbf{R}^d$ and $i = 0, \ldots, n$ respectively.

Theorem 6.4

$$\text{span}\{A_i(x), [A_j, A_k](x), [[A_j, A_k], A_l](x), \ldots$$
$$1 \le i \le n, \qquad 0 \le j, k, l, \ldots \le n\} \subseteq \text{Range } \sigma_t \quad a.s.$$

Proof For each $0 < s < t$, define

$$R_s = \text{span}\{Z(u)A_i(\xi_u) : 0 \le u \le s, \qquad i = 1, \ldots, n\}$$

and

$$R = R(\omega) = \bigcap_{s>0} R_s.$$

Then one can easily show that $R_t = \text{Range}\, \sigma_t$. Furthermore by the Blumenthal 0–1 law there exists a (deterministic) set $\tilde{R}$ such that $R(\omega) = \tilde{R}$ a.s. Suppose that $y \in \tilde{R}^{\perp}$. Then with probability 1 there exists $\tau > 0$ such that $R_s = \tilde{R}$ for $s \in [0, \tau]$. We then have

$$\langle Z(s)A_i(\xi_s), y \rangle = 0, \qquad s \in [0, \tau]$$

for each $i = 1, \ldots, n$. Iterating Lemma 6.3 on this gives

$$\langle Z(s)A_i(\xi_s), y \rangle, \langle Z(s)[A_j, A_k](\xi_s), y \rangle,$$
$$\langle Z(s)[[A_j, A_k], A_l](\xi_s), y \rangle, \ldots = 0 \quad \text{on} \quad [0, \tau]$$

for all $i = 1, \ldots, n; j, k, l, \ldots = 0, \ldots, n$. Evaluating this at $s = 0$ shows that $y \in (\text{span}\, H)^{\perp}$ where H is the set of vectors listed in (6.1). Thus $H \subseteq \tilde{R} \subseteq R_t = \text{Range}\, \sigma_t$ a.s. as required. ∎

Corollary *σ_t is invertible with probability* 1.

6.2 Hörmander's condition implies $\sigma^{-1} \in L^p$ for all p

Recall that we need to show this in order to complete the proof of Theorem 6.1. Although the result was first obtained by Kusuoka and Stroock[18], the proof that we shall give here is due to Norris[24]. Norris's work follows the same lines as that of Kusuoka and Stroock; however, his discovery of the following elementary proof of the next lemma (which constitutes a key step) considerably shortens their original argument.

Lemma 6.5 *Let $x_0, y_0 \in \mathbf{R}$ and suppose that a, $b = (b_1, \ldots, b_n)$ and $v = (v_1, \ldots, v_n)$ are adapted processes. Define*

$$x_t \equiv x_0 + \int_0^t a_s \, ds + \sum_{i=1}^n \int_0^t b_i(s) \, dw_i(s)$$

and

$$y_t \equiv y_0 + \int_0^t x_s \, ds + \sum_{i=1}^n \int_0^t v_i(s) \, dw_i(s).$$

Suppose further that $T \le t_0$ is a bounded stopping time and C is a constant such that

$$|a_t|, \|b_t\|, |x_t|, \|v_t\| \le C \qquad \text{for all } t \le T.$$

Then for each $q > 17$, there exist positive constants K, c and d such that

$$P\left(\int_0^T y_t^2 \, dt < \varepsilon^q \quad \text{and} \quad \int_0^T (x_t^2 + \|v_t\|^2) \, dt \ge \varepsilon\right)$$
$$\le K \exp(-c/\varepsilon) \quad \text{for all} \quad \varepsilon \in (0, d).$$

Proof Make the following definitions:

$$X_t \equiv \int_0^t x_s \, ds, \qquad M_t \equiv \sum_{i=1}^n \int_0^t v_i(s) \, dw_i(s),$$

$$N_t \equiv \sum_{i=1}^n \int_0^t y_s v_i(s) \, dw_i(s), \qquad Q_t \equiv \sum_{i=1}^n \int_0^t X_s b_i(s) \, dw_i(s).$$

For each ε, $\delta > 0$, define

$$B_1(\varepsilon, \delta) \equiv \left\{ \int_0^T y_s^2 \, \|v(s)\|^2 \, ds < \varepsilon \quad \text{and} \quad \sup_{t \leq T} |N_t| \geq \delta \right\}$$

$$B_2(\varepsilon, \delta) \equiv \left\{ \int_0^T \|v(s)\|^2 < \varepsilon \quad \text{and} \quad \sup_{t \leq T} |M_t| \geq \delta \right\}$$

$$B_3(\varepsilon, \delta) \equiv \left\{ \int_0^T X_s^2 \, \|b(s)\|^2 \, ds < \varepsilon \quad \text{and} \quad \sup_{t \leq T} |Q_t| \geq \delta \right\}.$$

It follows from a standard martingale inequality that

$$P(B_i(\varepsilon, \delta)) \leq 2e^{-\delta^2/2\varepsilon} \quad \text{for} \quad i = 1, 2, 3.$$

Let $q_1 = 1/2(q-1)$, $q_2 = 1/8(q-5)$, $q_3 = 1/8(q-9)$ and note that $q_3 > 1$. For $i = 1, 2, 3$, let $\delta_i = \varepsilon^{q_i}$. Positive constants ε_1, ε_2, ε_3 will now be chosen so that the following conditions are satisfied for some positive constant c:

$$\text{for} \quad i = 1, 2, 3, \; P(B_i) < 2e^{-c/\varepsilon} \quad \text{for all } \varepsilon \quad \text{where} \quad B_i \equiv B_i(\varepsilon_i, \delta_i) \tag{6.4}$$

$$\left\{ \int_0^T y_t^2 < \varepsilon^q \quad \text{and} \quad \int_0^T (x_t^2 + \|v_t\|^2) \, dt \geq \varepsilon \right\} \subseteq B_1 \cup B_2 \cup B_3 \qquad \text{for all } \varepsilon \text{ in some interval } (0, d). \tag{6.5}$$

This will obviously suffice to prove the lemma. Choose $\varepsilon_1 = C^2 \varepsilon^q$. It is easy to check that ε_1 satisfies (6.4) as do ε_2 and ε_3 which will be chosen later. Suppose that

$$\int_0^T y_t^2 \, dt < \varepsilon^q.$$

Then

$$\int_0^T y_t^2 \, \|v_t\|^2 \, dt < C^2 \varepsilon^q.$$

Suppose further that $\omega \notin B_1$. Then

$$\sup_{t \le T} \left| \sum_{i=1}^{n} \int_0^t y_s v_i(s)\, dw_i(s) \right| \le \delta = \varepsilon^{q_1}.$$

Also

$$\sup_{t \le T} \left| \int_0^t y_s x_s\, ds \right| \le \left(t_0 \int_0^T y_s^2 x_s^2\, ds \right)^{1/2} \le t_0^{1/2} C \varepsilon^{q/2}.$$

Since

$$dy_s = x_s\, ds + \sum_{i=1}^{n} v_i(s)\, dw_i(s)$$

it follows that

$$\sup_{t \le T} \left| \int_0^t y_s\, dy_s \right| < (1 + t_0^{1/2} C \varepsilon^{1/2}) \varepsilon^{q_1}.$$

However, by Itô's lemma

$$y_t^2 = y_0^2 + 2 \int_0^t y_s\, dy_s + \int_0^t \|v(s)\|^2\, ds,$$

so

$$\int_0^T \int_0^t \|v(s)\|^2\, ds = \int_0^T y_t^2\, dt - T y_0^2$$
$$- 2 \int_0^T \left(\int_0^t y_s\, dy_s \right) dt < \varepsilon^q + 2t_0 (1 + t_0^{1/2} C \varepsilon^{1/2}) \varepsilon^{q_1}$$

This implies

$$\int_0^T \int_0^t \|v(s)\|^2\, ds\, dt < (2t_0 + 1) \varepsilon^{q_1} \quad \text{for sufficiently small } \varepsilon. \tag{6.6}$$

Suppose now that f is a positive increasing function on $[0, T]$ with $\int_0^T f(s)\, ds = \alpha$. Then for any $t \in (0, T]$

$$\alpha = \int_0^T f(T - s)\, ds \ge \int_0^t f(T - s)\, ds \ge \int_0^t f(T - t)\, ds = t f(T - t).$$

Thus $f(T - t) \le \alpha / t$. Applying this in (6.6) gives

$$\int_0^{T-t} \|v(s)\|^2\, ds \le (2t_0 + 1) \varepsilon^{q_1} / t$$

which implies

$$\int_0^T \|v(s)\|^2\, ds \le (2t_0 + 1) \varepsilon^{q_1} / t + C^2 t \tag{6.7}$$

Choose

$$\varepsilon_2 = (1 + C^2)(2t_0 + 1)^{1/2}\varepsilon^{q_1/2}.$$

Applying (6.7) with

$$t = (2t_0 + 1)^{1/2}\varepsilon^{q_1/2}$$

shows that

$$\int_0^T \|v(s)\|^2 \, ds \le \varepsilon_2.$$

Assume that $\omega \notin B_2$. Then $\sup_{t \le T} |M_t| < \delta_2 = \varepsilon^{q_2}$. In view of the fact that

$$\int_0^T y_t^2 \, dt < \varepsilon^q$$

we have

$$\lambda\{t \in [0, T] : |y_t| \ge \varepsilon^{q/3}\} \le \varepsilon^{q/3}$$

where λ denotes Lebesgue measure. Hence

$$\lambda\{t \in [0, T] : |y_0 + X_t| \ge \varepsilon^{q/3} + \varepsilon^{q_2}\} \le \varepsilon^{q/3}.$$

So for each $t \in [0, T]$ there exists $s \in [0, T]$ such that

$$|s - t| \le \varepsilon^{q/3}$$

and

$$|y_0 + X_s| < \varepsilon^{q/3} + \varepsilon^{q_2}.$$

Hence

$$|y_0 + X_t| \le |y_0 + X_s| + \int_s^t |x_r| \, dr \le (1 + C)\varepsilon^{q/3} + \varepsilon^{q_2}.$$

Taking $t = 0$ gives $|y_0| \le (1 + C)\varepsilon^{q/3} + \varepsilon^{q_2}$, so

$$|X_t| < 2\{(1 + C)\varepsilon^{q/3} + \varepsilon^{q_2}\} \le 3\varepsilon^{q_2}$$

for sufficiently small ε.

Applying Itô's lemma, we have

$$\int_0^T x_t^2 \, dt = \int_0^T x_t \, dX_t = x_T X_T - \int_0^T X_s a_s \, ds - \sum_{i=1}^n \int_0^T b_i(s) \, dw_i(s) \tag{6.8}$$

However

$$|x_T X_T| < 3C\varepsilon^{q_2} \tag{6.9}$$

$$\left|\int_0^T X_t a_t \, dt\right| < 3Ct_0\varepsilon^{q_2} \tag{6.10}$$

and

$$\int_0^T X_t^2 \, \|b(t)\|^2 \, dt \leq 9C^2 t_0 \varepsilon^{2q_2}. \tag{6.11}$$

Choose $\varepsilon_3 = 9C^2 t_0 \varepsilon^{2q_2}$ and suppose that $\omega \notin B_3$. Then (6.11) implies

$$|Q_T| = \left|\sum_{i=1}^n \int_0^T X_t b_i(t) \, dw_i(t)\right| < \delta_3 = \varepsilon^{q_3} \tag{6.12}$$

(6.8)–(6.10) and (6.12) give

$$\int_0^T x_t^2 \, dt < 3C(1 + t_0)\varepsilon^{q_2} + \varepsilon^{q_3} \leq 2\varepsilon^{q_3}.$$

Together with (6.7) (applied with $t = \varepsilon^{q_1/2}$) this gives

$$\int_0^T (x_t^2 + \|v(t)\|^2) \, dt < 2\varepsilon^{q_3} + \varepsilon^{q_1/2}(C^2 + 2t_0 + 1) < \varepsilon$$

for ε small enough.

It has been shown that if ε is small enough, $\omega \notin B_1 \cup B_2 \cup B_3$ and $\int_0^T y_t^2 \, dt < \varepsilon^q$, then $\int_0^T (x_t^2 + \|v(t)\|^2) \, dt < \varepsilon$. Thus (6.5) holds and the proof of the lemma is complete. ■

The proof of the next result is elementary.

Lemma 6.6 *Suppose that X is a random variable such that*

$$P(X < \varepsilon) = 0(\varepsilon^k) \quad \text{as} \quad \varepsilon \to 0 \quad \text{for every} \quad k \in \mathbf{N}.$$

Then $X^{-1} \in \bigcap_{p=1}^{\infty} L^p$.

We now introduce some more notation. For every $\ell \in \mathbf{N}$, denote by K_ℓ the set of vector fields appearing in (6.1) which contain at most $(\ell - 1)$ iterated Lie brackets. Choose and fix an integer ℓ such that K_ℓ spans $\mathbf{R}^d$ and set

$$\delta = \inf_{\|v\|=1} \left\{ \sup_{K \in K_\ell} \langle K(x), v\rangle^2 \right\}.$$

Note that since the infimum is taken over a compact set, δ is strictly positive. Let S denote the unit circle in $\mathbf{R}^d$. Finally, given a number B, define a stopping time T such that

$$T = \inf\{s \geq 0 : \|\xi_s - x\| \geq 1/B \quad \text{or} \quad \|Z_s - I\| \geq 1/B\} \wedge t.$$

Then

$$\{T \leq \varepsilon\} = \left\{\sup_{s \leq \varepsilon} \|\xi_s - x\| \vee \sup_{s \leq \varepsilon} \|Z_s - I\| \geq 1/B\right\}. \tag{6.13}$$

It follows from Theorem 1.9(iv) that

$$E\left\{\sup_{s \leq \varepsilon} \|\xi_s - x\|^p \vee \sup_{s \leq \varepsilon} \|Z_s - I\|^p\right\} = 0(\varepsilon^{p/2}) \quad \text{for all } p. \tag{6.14}$$

Equations (6.13) and (6.14) imply that $P(T \leq \varepsilon) = 0(\varepsilon^p)$ for all p, so by Lemma 6.6

$$T^{-1} \in L^p \quad \text{for all } p. \tag{6.15}$$

Since all the functions K contained in K_ℓ are continuous, the constant B which appears in the definition of T may (and will) be chosen large enough to ensure that: for each $v \in S$, there exist $K \in K_\ell$ and a neighbourhood N of v in S such that

$$\inf_{s \leq T, u \in N} \langle Z_s K(\xi_s), u\rangle \geq \delta/2. \tag{6.16}$$

This condition together with (6.15) implies: for each $v \in S$ there exist $K \in K_\ell$ and a neighbourhood N of v in S such that for all P

$$\sup_{u \in N} P\left(\int_0^T \langle Z_s K(\xi_s), u\rangle^2 \, ds < \varepsilon\right) \leq P\left(\frac{\delta T}{2} < \varepsilon\right) = 0(\varepsilon^P). \tag{6.17}$$

The remainder of the proof is structured as follows. Recall that

$$\sigma = \int_0^t Z_s A(\xi_s) A(\xi_s)^* Z_s^* \, ds,$$

where A is bounded and Z is the solution of a linear stochastic differential equation with bounded coefficients. Theorem 1.9(iv) implies that for all $i, j \in \{1, \ldots, d\}$,

$$\sigma_{ij} \in \bigcap_{p=1}^{\infty} L^p.$$

Hence it will suffice to show that

$$\det \sigma^{-1} \in \bigcap_{p=1}^{\infty} L^p. \tag{6.18}$$

Denote by λ the smallest eigenvalue of σ. Showing that $\lambda^{-1} \in \bigcap_{p=1}^{\infty} L^p$, will prove (6.18). By Lemma 6.6 this is implied by the statement

$$P(\lambda < \varepsilon) = 0(\varepsilon^p) \qquad \text{for all } p. \tag{6.19}$$

In Lemma 6.8 we will prove the following result, from which (6.19) follows as an immediate consequence

$$P\left(\inf_{v \in S} \sum_{i=1}^{n} \int_0^T \langle Z_s A_i(\xi_s), v\rangle^2 \, ds < \varepsilon\right) = 0(\varepsilon^p) \qquad \text{for all } p. \tag{6.20}$$

Note that in (6.17) we have a statement somewhat similar to this. The differences between (6.17) and (6.20) are:

(i) the presence of the infimum inside the probability in (6.20);

(ii) the fact that in (6.17) we have an assertion about an arbitrary element K in K_ℓ, whereas in (6.20) we require this statement for (at least) one of the particular vector fields $A_1, \ldots, A_n$.

In the sequel the former discrepancy will be taken care of by a (quite easy) compactness argument and the latter by an inductive argument which turns on Lemma 6.5.

Lemma 6.7 *For every $v \in S$ there exist $i \in \{1, \ldots, n\}$ and a neighbourhood N of v in S such that*

$$\sup_{u \in N} P\left(\int_0^T \langle Z_s A_i(\xi_s), u\rangle^2 \, ds < \varepsilon\right) = 0(\varepsilon^p) \quad \text{for all } p.$$

Proof Without loss of generality one may assume that the vector field K in (6.17) is of the form

$$K = \pm[[A_{i_1}, A_{i_2}], A_{i_3} \ldots], A_{i_r}]$$

where $r \leq \ell$, $i_1, \ldots, i_r \in \{0, 1, \ldots, n\}$ and $i_1 \neq 0$. Define $K_1 = A_{i_1}$ and $K_j = [K_{j-1}, A_{i_j}]$, $j = 2, \ldots, r$. It will be shown by induction on j (decreasing) that for $j = 1, \ldots, r$

$$\sup_{u \in N} P\left(\int_0^T \langle Z_s K_j(\xi_s), u\rangle^2 \, ds < \varepsilon\right) = 0(\varepsilon^p) \qquad \text{for all } p \tag{6.21}$$

which will suffice to prove the result.

Itô's lemma gives

$$d(Z_s K_{j-1}(\xi_s)) = \sum_{i=1}^{n} Z_s[A_i, K_{j-1}](\xi_s)\, dw_i(s)$$
$$+ Z_s\left\{[A_0, K_{j-1}](\xi_s) + \frac{1}{2}\sum_{i=1}^{n} [[A_i, K_{j-1}], A_i](\xi_s)\right\} ds$$

Define $y_s = \langle Z_s K_{j-1}(\xi_s), u\rangle$

$$x_s = \left\langle Z_s[A_0, K_{j-1}](\xi_s) + \frac{1}{2}\sum_{i=1}^{n} [[A_i, K_{j-1}], A_i](\xi_s), u\right\rangle$$

and $u_i(s) = \langle Z_s[A_i, K_{j-1}](\xi_s), u\rangle$; $i = 1, \ldots, n$.

The definition of T ensures that the processes y, x and $u = (u_1, \ldots, u_n)$ satisfy the hypotheses of Lemma 6.5 on $[0, T]$ with the constant C independent of $u \in N$. Applying Lemma 6.5 gives

$$P\left(\int_0^T \langle Z_s K_{j-1}(\xi_s), u\rangle^2 \, ds < \varepsilon^{18}\right.$$

and

$$\int_0^T \left\langle Z_s\left\{[A_0, K_{j-1}](\xi_s) + \frac{1}{2}\sum_{i=1}^{n} [[A_i, K_{j-1}], A_i](\xi_s)\right\}, u\right\rangle^2$$

$$\left. + \sum_{i=1}^{n} \langle Z_s[A_i, K_{j-1}](\xi_s), u\rangle^2 \, ds \geq \varepsilon\right) = 0(\varepsilon^p) \tag{6.22}$$

for all p uniformly in $u \in N$. If $i_j \neq 0$, then K_j is one of the vector fields $[A_i, K_{j-1}]$, $i = 1, \ldots, n$ and it follows from the elementary inequality

$$P(\Omega_1) \leq P(\Omega_1 \cap \Omega_2) + P(\Omega_2^c)$$

that (6.21) holds for $j - 1$.

Suppose, however, that $i_j = 0$. Write

$$\Omega = \left\{\int_0^T \langle Z_s K_{j-1}(\xi_s), u\rangle^2 \, ds < \varepsilon^{18}\right\}$$

$$W_s = \langle Z_s[A_0, K_{j-1}](\xi_s), u\rangle$$

and

$$Y_i(s) = \langle Z_s[[A_i, K_{j-1}], A_i](\xi_s), u\rangle \qquad i = 1, \ldots, n.$$

Repeating the above argument with $[K_{j-1}, A_i]$ in place of K_{j-1} (for each $i = 1, \ldots, n$) gives

$$P\left(\Omega \cap \left\{\int_0^T Y_i^2(s) \geq \varepsilon\right\}\right) = 0(\varepsilon^p) \qquad \text{for all } p. \tag{6.23}$$

The required result will follow from (6.22) if we show

$$P\left(\Omega \cap \left\{\int_0^T \left(W_s + \frac{1}{2}\sum_{i=1}^{n} Y_i(s)\right)^2 ds < \varepsilon\right\}\right) = 0(\varepsilon^p) \quad \text{for all } p. \tag{6.24}$$

However

$$P\left(\Omega \cap \left\{\int_0^T \left(W_s + \frac{1}{2}\sum_{i=1}^n Y_i(s)\right)^2 ds < \varepsilon\right\}\right)$$
$$\leq P\left(\Omega \cap \left\{\int_0^T \left(W_s + \frac{1}{2}\sum_{i=1}^n Y_i(s)\right)^2 ds < \varepsilon\right\}\right.$$
$$\left.\cap \bigcap_{i=1}^n \left\{\int_0^T Y_i^2(s)\, ds < \varepsilon\right\}\right) + \sum_{i=1}^n P\left(\int_0^T Y_i^2(s)\, ds > \varepsilon\right).$$

By (6.23) the last term is $0(\varepsilon^p)$ for all p, while

$$P\left(\Omega \cap \left\{\int_0^T \left(W_s + \frac{1}{2}\sum_{i=1}^n Y_i(s)\right)^2 ds < \varepsilon\right\}\right.$$
$$\left.\cap \bigcap_{i=1}^n \left\{\int_0^T Y_i(s)^2\, ds < \varepsilon\right\}\right)$$
$$\leq P\left(\int_0^T W_s^2\, ds \leq \varepsilon(2 + n2^{n-1})\right) = 0(\varepsilon^p) \qquad \text{for all } p \text{ by (6.21).}$$

Thus (6.24) holds and the inductive step is complete. ■

Remark Note that the final stages in the proofs of the last result and Lemma 6.3 are similar in that they both consist of separating the vector fields

$$Z_s[A_0, K_{j-1}](\xi_s) \quad \text{and} \quad \sum_{i=1}^n [[A_i, K_{j-1}], A_i](\xi_s).$$

Lemma 6.8

$$P\left(\inf_{v \in S} \sum_{i=1}^n \int_0^T \langle Z_s A_i(\xi_s), v\rangle^2\, ds < \varepsilon\right) = 0(\varepsilon^p) \quad \textit{for all } p.$$

Proof By choice of T the quadratic forms

$$v \to \int_0^T \sum_{i=1}^n \langle Z_s A_i(\xi_s), v\rangle^2\, ds$$

are uniformly Lipschitz on S.

Let θ denote their common Lipschitz constant and cover S with balls of radius ε/θ and centre v_j. The number of these balls may be chosen to be less than $D(\theta/\varepsilon)^d$ for some fixed number D. Then

$$\int_0^T \sum_{i=1}^n \langle Z_s A_i(\xi_s), v\rangle^2\, ds < \varepsilon \quad \text{for some } v \in S \text{ implies}$$

$$\int_0^T \sum_{i=1}^n \langle Z_s A_i(\xi_s), v_j\rangle^2\, ds < 2\varepsilon \quad \text{for some } j.$$

Hence

$$P\left(\inf_{v\in S} \int_0^T \sum_{i=1}^n \langle Z_s A_i(\xi_s), v\rangle^2\, ds < \varepsilon\right)$$
$$\leq D(\theta/\varepsilon)^d \sup_j P\left(\int_0^T \sum_{i=1}^n \langle Z_s A_i(\xi_s), v_j\rangle^2\, ds < 2\varepsilon\right)$$
$$\leq D(\theta/\varepsilon)^d \sup_{v\in S} P\left(\int_0^T \sum_{i=1}^n \langle Z_s A_i(\xi_s), v\rangle^2\, ds < 2\varepsilon\right).$$

Since S is compact, Lemma 6.7 implies that this last term is $0(\varepsilon^p)$ for all p and this completes the proof. ■

7

Some further applications of the Malliavin calculus

7.1 The filtering problem

The filtering problem may be formulated in the following way. Suppose that it is required to determine the position of an object (e.g. a spacecraft) at a certain time, and this depends upon two factors A and B, where only B is observable.† Then under certain assumptions A and B may be modelled as the solutions $\{\xi_t : t \geq 0\}$ and $\{\theta_t : t \geq 0\}$ of stochastic differential equations and the position of the object can be expressed as a progressively measurable function h of ξ and θ. In view of the fact that ξ is unknown, the best estimate for the position at any time t will be given by the conditional expectation

$$E[h(t, \xi, \theta)/\theta_s : s \leq t]. \tag{7.1}$$

Now it turns out that if the conditional probabilities $P(\xi_t \in \mathrm{d}x/\theta_s : s \leq t)$ are absolutely continuous with respect to the Lebesgue measure, then a relatively simple stochastic differential equation can be obtained for (7.1). More generally an equation of this simple type is not available, and then the analysis of (7.1) is considerably more difficult. Thus an important problem in filtering theory is the investigation of when the random variables ξ_t, conditioned on the past history of the process θ, have absolutely continuous distributions. The present section is a brief account of the treatment of this problem given by Dominique Michel[22].

The basic notations and hypotheses are as follows: suppose that T is a fixed positive time and n, p and d are natural numbers such that $n = d + p$. Let A, B, a and b be functions satisfying the following

† A and B generally involve noise and are therefore non-deterministic.

conditions:

(i) $A\colon [0, T] \times \mathbf{R}^d \times C(\mathbf{R}^p) \to \mathbf{M}^{n,d}$

$B\colon [0, T] \times \mathbf{R}^d \times C(\mathbf{R}^p) \to \mathbf{R}^d$

$a\colon [0, T] \times C(\mathbf{R}^p) \to \mathbf{M}^{n,p}$

$b\colon [0, T] \times \mathbf{R}^d \times C(\mathbf{R}^p) \to \mathbf{R}^p$.

Here $C(\mathbf{R}^p)$ denotes the set of paths from $[0, T]$ into $\mathbf{R}^p$.

(ii) Let f be any one of the functions in (i). Then for each appropriate pair (t, x) the map on $C(\mathbf{R}^p)$ defined by $c \to f(t, x, c)$ depends only on $\{c(s) : s \leq t\}$. Furthermore there exists an increasing function K, $0 < K < 1$ and constants L_1 and L_2 such that

$$\|f(t, x, c) - f(t, x', c')\| \leq L_1 \int_0^t \|c_s - c_s'\|^2 \,\mathrm{d}K(s) + L_2\{\|x - x'\|^2 + \|c_t - c_t'\|^2\}$$

and

$$f^2(t, x, c) \leq L_1 \int_0^t (1 + \|c_s\|^2) \,\mathrm{d}K(s) + L_2(1 + \|x\|^2 + \|c_t\|^2).$$

(iii) There exists a constant $\varepsilon > 0$ such that

$$\langle aa^*z, z\rangle \geq \varepsilon \|z\|^2 \,\forall\, z \in \mathbf{R}^p$$
$$\langle \tilde{A}z, z\rangle \geq \varepsilon \|z\|^2 \,\forall\, z \in \mathbf{R}^d,$$

where $\tilde{A} \equiv A(I_n - a^*(aa^*)^{-1}a)A^*$.

(iv) The functions A, B, a, b are all bounded. Furthermore each of these maps is thrice differentiable in the second argument and its corresponding partial derivatives in this argument are bounded up to third order.

Let ξ_0 and θ_0 be random variables in $\cap_{p=1}^{\infty} L^p$ and define (ξ, θ) to be the solution of the stochastic differential system

$$\xi_t = \xi_0 + \int_0^t A(s, \xi_s, \theta) \,\mathrm{d}w_s + \int_0^t B(s, \xi_s, \theta) \,\mathrm{d}s$$
$$t \in [0, T]$$
$$\theta_t = \theta_0 + \int_0^t a(s, \theta) \,\mathrm{d}w_s + \int_0^t b(s, \xi_s, \theta) \,\mathrm{d}s$$

where, as before, w denotes Brownian motion in $\mathbf{R}^n$. Let $\mathscr{F}_t = \mathscr{F}\{w_s : s \leq t\}$ and $\mathscr{F}_t^\theta = \mathscr{F}\{\theta_s : s \leq t\}$ for each $t \in [0, t]$.

The above system may be reformulated as

$$\mathrm{d}\xi_t = \tilde{A}(t, \xi_t, \theta) \,\mathrm{d}\tilde{w}_t^1 + \tilde{B}(t, \xi_t, \theta) \,\mathrm{d}\theta_t + C(t, \xi_t, \theta) \,\mathrm{d}t \tag{7.2}$$
$$\mathrm{d}\theta_t = aa^*(t, \theta) \,\mathrm{d}\tilde{w}_t^2 + b(t, \xi_t, \theta) \,\mathrm{d}t$$

where $\tilde{w} = (\tilde{w}_1, \ldots, \tilde{w}_n)$ is another n-dimensional Brownian motion, $\tilde{w}^2 = (\tilde{w}_1, \ldots, \tilde{w}_p)$, $\tilde{w}^1 = (\tilde{w}_{p+1}, \ldots, \tilde{w}_n)$, and $\tilde{B}$ and C are maps from $[0, T] \times \mathbf{R}^d \times C(\mathbf{R}^p)$ into $\mathbf{M}^{p,d}$, $\mathbf{R}^d$ and $\mathbf{M}^{p,p}$ respectively, which are defined in terms of A, B, a and b. It is shown that for each t, equation (7.2) defines ξ_t as a Borel function of $(\xi_0, \tilde{w}^1, \theta)$ with respect to the product field on $\mathbf{R}^d \times C(\mathbf{R}^d \times \mathbf{R}^d)$. The solution of (7.2) with initial point x will be denoted by ξ^x (or by $\xi^x(\tilde{w}^1)$ when we wish to emphasize its dependence on $\tilde{w}^1$).

The following result reduces the analysis of the conditional distribution of ξ_t to a problem of integration by parts on Wiener space. It is a functional analogue of Bayes formula, obtained from the Girsanov theorem.

Theorem 7.1 *Suppose that φ is a bounded real-valued function on $\mathbf{R}^d$. Then for every $t \in [0, T]$*

$$E[\varphi(\xi_t)/\mathscr{F}_t^\theta] = \int_{\mathbf{R}^d \times C_0} \varphi(\xi_t^x(w))\rho_t(x, w, \theta)\, \mathrm{d}F(x)\, \mathrm{d}\gamma(w) \tag{7.3}$$

where $\mathrm{d}F(x) = P(\xi_0 \in \mathrm{d}x/\theta_0)$, γ is the Wiener measure on C_0 $(= C_0(\mathbf{R}^d))$, and $\rho_t(x, w, \theta)$ is the solution of the stochastic differential equation

$$\mathrm{d}\rho_t(x, w, \theta) = \rho_t(x, w, \theta)[b(t, \xi_t^x, \theta) - \bar{b}(t, \theta)]^*(aa^*)(t, \theta)^{-1/2}\, \mathrm{d}\theta_s$$

$$\rho_0(x, w, \theta) = 1$$

where $\bar{b}(s, \theta) = E[b(s, \xi_s, \theta)/\mathscr{F}_s^\theta]$.

It follows as an immediate corollary that if φ is any test function on $\mathbf{R}^d$ and $y \in \mathbf{R}^d$, then

$$E[D\varphi(\xi_t)y/\mathscr{F}_t^\theta] = \int_{\mathbf{R}^d} I_t(x, \theta)\, \mathrm{d}F(x) \tag{7.4}$$

where

$$I_t(x, \theta) = \int_{C_0} D\varphi(\xi_t^x)y\rho_t(x, w, \theta)\, \mathrm{d}\gamma(w). \tag{7.5}$$

Let $\mathscr{L}$ and $\langle , \rangle$ be the linear operator and bilinear form introduced in chapter 2, acting on the d-dimensional Wiener space C_0. As before, let $\mathscr{D}(\mathscr{L})$ denote the domain of $\mathscr{L}$ and $\mathscr{K}_p$ the subset of $\mathscr{D}(\mathscr{L})$ containing functionals f such that

$$\|f\|_{\mathscr{K}}^{2p} \equiv E[f^{2p} + (\mathscr{L}f)^{2p} + (\langle f, f \rangle)^p] < \infty.$$

Define $\mathscr{K} = \bigcap_{p=1}^\infty \mathscr{K}_p$ and for each p define $\mathscr{K}_{p,T}$ to be the set of non-anticipating processes α into $\mathcal{K}_p$ such that

$$\|\alpha\|^{2p}_{\mathcal{K}_{p,T}} = E \sup_{t \le T} [\alpha_t^{2p} + (\mathcal{L}\alpha_t)^{2p} + (\langle \alpha_t, \alpha_t \rangle)^p] < \infty.$$

Finally let $\mathcal{K}_T = \cap_{p=1}^{\infty} \mathcal{K}_{p,T}$. Note that $\mathcal{K}$ and $\mathcal{K}_T$ form an algebra under pointwise multiplication and addition.

The main result of this section is:

Theorem 7.2 (Michel) *The conditional measure $P(\xi_t \in \mathrm{d}x / \mathcal{F}_t^\theta)$ is absolutely continuous with respect to the Lebesgue measure on $\mathbf{R}^d$, almost surely with respect to the law of θ, for all $t \in (0, T]$.*

The theorem will be proved by effecting an integration by parts in (7.5). Since the integrand in (7.5) involves functionals of a more general type than those considered in chapter 2, this step requires the following result.

Theorem 7.3 *Let D, E, F be maps from $[0, \infty) \times \mathbf{R}^r \times \mathbf{R}^s \times C(\mathbf{R}^p)$ into $\mathbf{M}^{d,s}$, $\mathbf{M}^{p,s}$ and $\mathbf{R}^s$ respectively satisfying hypotheses* (ii) *on page* 88. *Assume further that these maps are C^2 with respect to their second and third arguments with bounded first and second derivatives. Let η be a continuous process with values in $\mathbf{R}^r$ such that all of its components are in $\mathcal{K}_T$. Let X_0 be an $\mathcal{F}_0$ measurable random variable in L^p for all p. Then the equation*

$$X_t = X_0 + \int_0^t D(s, \eta_s, X_s, \theta)\, \mathrm{d}w_s + \int_0^t E(s, \eta_s, X_s, \theta)\, \mathrm{d}\theta_s$$

$$+ \int_0^t F(s, \eta_s, X_s, \theta)\, \mathrm{d}s; \qquad t \in [0, T]$$

has a unique solution. Furthermore, $X_t \in \mathcal{K}$ for all $t \in [0, T]$ and $\mathcal{L}X_t$ and $\langle X_i(t), X_j(t) \rangle$ for $1 \le i, j \le s$ satisfy

$$\mathcal{L}X_t = \int_0^t (\mathcal{L}D - D/2)_s\, \mathrm{d}w_s + \int_0^t (\mathcal{L}E)_s\, \mathrm{d}\theta_s + \int_0^t (\mathcal{L}F)_s\, \mathrm{d}s$$

$$\langle X_i(t), X_j(t) \rangle = \int_0^t [\langle X_i(s), D_j(s) \rangle + \langle X_j(s), D_i(s) \rangle]\, \mathrm{d}w_s$$

$$+ \int_0^t [\langle X_i(s), E_j(s) \rangle + \langle X_j(s), E_i(s) \rangle]\, \mathrm{d}\theta_s$$

$$+\int_0^t [\langle X_i(s), F_j(s)\rangle + \langle X_j(s), F_i(s)\rangle$$

$$+\sum_k \langle D_{ki}(s), D_{kj}(s)\rangle + \langle (Ea)_{ki}(s), (a^*E^*)_{kj}(s)\rangle]\, ds.$$

The proof uses an iterative procedure similar to that of Theorem 2.10.

It follows from Theorem 7.3 that for each $x \in \mathbf{R}^d$, $t \in [0, T]$ and $1 \le i \le d$, the map $w \to \xi_i^x(t)$ is in $\mathscr{D}(\mathscr{L})$, and the same is true of the map $w \to \rho_t(x, w, \theta)$ where $\rho_t(x, w, \theta)$ is as in (7.3). Let σ_t^x denote the $d \times d$ matrix $(\langle \xi_i^x(t), \xi_j^x(t)\rangle)_{i,j=1}^d$ and Δ_t^x the determinant of σ_t^x. It is shown that the hypothesis $(\langle \tilde{A}z, z\rangle \ge \varepsilon \|z\|^2 \,\forall\, z \in \mathbf{R}^d)$ implies that Δ_t^x is almost surely non-vanishing and $(\Delta_t^x)^{-1} \in \mathscr{K}$. Finally Theorem 7.3 implies that $\rho_t(x, w, \theta) \in \mathscr{K}$. The proof of Theorem 7.2 now proceeds along the same lines as that of Theorem 2.5. Let $(P_{jk}^x(t))_{j,k=1}^d$ denote the cofactor matrix of σ_t^x. If φ and y are as in (7.5), then one has

$$D\varphi(\xi_t^x)y = \frac{1}{\Delta_t^x} \sum_{j,k=1}^d \langle \varphi \circ \xi_t^x, \xi_j^x(t)\rangle P_{jk}^x(t) y_j$$

where $y_1, \ldots, y_d$ are the components of y with respect to any orthonormal basis of $\mathbf{R}^d$. Substituting this into (7.5), writing $\langle , \rangle$ in terms of $\mathscr{L}$ and then using the symmetry of $\mathscr{L}$ gives

$$\begin{aligned} I_t(x, \theta) = \sum_{j,k=1}^d y_j \int_{C_0} \varphi \circ \xi_t^x \Big\{ \xi_j^x(t) \mathscr{L}\Big[\frac{1}{\Delta_t^x} \rho_t(x, w, \theta) P_{jk}^x(t)\Big] \\ - \frac{1}{\Delta_t^x} \rho_t(x, w, \theta) P_{jk}^x(t) \mathscr{L}(\xi_j^x(t)) \\ - \mathscr{L}\Big[\frac{1}{\Delta_t^x} \rho_t(x, w, \theta) P_{jk}^x(t) \xi_j^x(t)\Big]\Big\}\, d\gamma(w). \end{aligned}$$

Using this relation in (7.4), together with the fact that θ_0 and ξ_0 are in L^P for all $P \in \mathbf{N}$, one obtains the estimate

$$|E[D\varphi(\xi_t)y / \mathscr{F}_t^\theta]| \le C \|\varphi\|_\infty$$

and the result follows from this.

Remark It is actually shown in Michel[22] that if A, B, a and b are infinitely differentiable in their second argument with bounded partial derivatives of all orders, then for each t the density function $P(\xi_t \in dx / \mathscr{F}_t^\theta)$ is smooth in x, almost surely with respect to the law of θ. As usual, the proof involves iterating the procedure used to prove the existence of the density.

7.2 A study of an infinite system of interacting particles

The following situation is of interest because it arises in statistical mechanics. Suppose that $\{P_k : k \in \mathbf{Z}\}$ is a denumerable collection of particles such that each P_k is acted upon by its near neighbours $P_{k-L}, \ldots, P_{k+L}$, where L is some fixed integer. Let $\xi_k(t)$ denote the position of the particle P_k at time t, for every $k \in \mathbf{Z}$ and $t \geq 0$. Then a natural way to study the distribution of the infinite dimensional process $\xi(\cdot) \equiv (\xi_k(\cdot))_{k \in \mathbf{N}}$ is via its finite dimensional projections $\xi^{(N)} \equiv (\xi_{-N}, \ldots, \xi_N)$, $N \in \mathbf{N}$. In particular one would like to show that under suitable conditions, the law of each process $\xi^{(N)}$ admits a density $F^{(N)}$ with respect to the Lebesgue measure on $\mathbf{R}^{2N}$, and that this density is smooth. It turns out to be sufficient to study this question with the infinite system above replaced by a finite one, as long as one can obtain estimates on the density functions $F^{(N)}$, $N \in \mathbf{N}$ which are independent of the cardinality of this finite system. In this section we will describe some work of Stroock[28] in which he uses the Malliavin calculus to obtain such estimates.† For further applications of these results the reader is referred to Holley and Stroock[13].

Let $A : \mathbf{R}^d \to \mathbf{M}^{d,d}$ and $B : \mathbf{R}^d \to \mathbf{R}^d$ be bounded, C^∞ functions with bounded derivatives of all orders. Suppose that there exists an integer L with $0 \leq L \leq d$ such that for all i, j and k

$$A_{ik} \equiv 0 \quad \text{if} \quad |i-k| > L \tag{7.6}$$

$$\frac{\partial A_{ik}}{\partial x_j} = \frac{\partial B_i}{\partial x_j} = 0 \quad \text{if} \quad |i-j| > L. \tag{7.7}$$

Assume that there exists $\varepsilon > 0$ such that for all y in $\mathbf{R}^d$

$$\langle AA^*(y)z, z\rangle \geq \varepsilon \, \|z\|^2 \quad \text{for all } z \text{ in } \mathbf{R}^d. \tag{7.8}$$

Let $x \in \mathbf{R}^d$ and denote by ξ the continuous adapted solution of the equation

$$\xi_t = x + \int_0^t A(\xi_s)\, \mathrm{d}w_s + \int_0^t B(\xi_s)\, \mathrm{d}s, \qquad t \geq 0$$

where w denotes Brownian motion in $\mathbf{R}^d$.

Finally let $0 \leq N \leq d$ be a fixed integer and denote by $\xi^{(N)}$ the process $(\xi_1, \ldots, \xi_N)$. The following result will be proved.

Theorem 7.4 *For every $t > 0$, $\xi^{(N)}(t)$ admits a smooth density F on $\mathbf{R}^N$. Furthermore for every $k \geq 0$ there exists a constant C, depending on*

† It should be noted that estimates of this kind cannot be obtained by classical techniques.

k, N and t but independent of $d \geq N$ such that

$$\|D^k F\|_\infty \leq C. \tag{7.9}$$

Let $\mathscr{L}$ and $\langle , \rangle$ be as in chapter 2, and $\mathscr{K}_p$, $\mathscr{K}$, $\mathscr{K}_{p,T}$ and $\mathscr{K}_T$ as in section 7.1. For any elements $\Phi_1, \ldots, \Phi_n \in \mathscr{K}$, let $\mathscr{G}(\Phi_1, \ldots, \Phi_n) = \{\Phi_1, \ldots, \Phi_n; \mathscr{L}\Phi_1, \ldots, \mathscr{L}\Phi_n; \langle \Phi_1, \Phi_1 \rangle^{1/2}, \ldots, \langle \Phi_n, \Phi_n \rangle^{1/2}\}$. Define $\mathscr{K}^{(n)}$ for $n \geq 1$ so that $\mathscr{K}^{(1)} = \mathscr{K}$ and $\mathscr{K}^{(n+1)} = \{\Phi \in \mathscr{K}^{(n)} : \mathscr{G}(\Phi) \subseteq \mathscr{K}\}$; where $\mathscr{G}^{(1)}(\Phi) = \{\Phi\}$ and $\mathscr{G}^{(n+1)} = \mathscr{G}(\mathscr{G}^{(n)}(\Phi))$. For $n \geq 1$ and $p \geq 1$, set

$$\|\Phi\|_{\mathscr{K}_p}^{(n)} = \left[\sum_{\psi \in \mathscr{G}^{(n)}(\Phi)} \|\psi\|_{\mathscr{K}_p}^{2p} \right]^{1/2p}.$$

Finally if α is an adapted process with values in $\mathscr{K}^p$ which satisfies $\alpha(t) \in \mathscr{K}^{(n)}$ for all $t \geq 0$, then define

$$\|\alpha(\cdot)\|_{\mathscr{K}_{p,T}}^{(n)} = \left[\sum_{\beta(\cdot) \in \mathscr{G}^{(n)}(\alpha(\cdot))} \|\beta(\cdot)\|_{\mathscr{K}_{p,T}}^{2p} \right]^{1/2p}.$$

The proof of Theorem 7.4 will require estimates on the moments of both $\xi_t^{(N)}$ and $(\det \sigma_t^{(N)})^{-1}$, which are independent of $d \geq N$, where $\sigma_t^{(N)}$ denotes the covariance matrix $(\langle \xi_i(t), \xi_j(t) \rangle)_{i,j=1}^N$. The next result furnishes estimates of the former kind.

Lemma 7.5 *For each $n \geq 1$, $p \geq 2$ and $T > 0$, there exists a constant $C(n, P, T) < \infty$, depending only on the bounds on A and B and their derivatives up to order n, such that*

$$\|\xi_i(\cdot)\|_{\mathscr{K}_{P,T}}^{(n)} \leq C(n, P, T)$$

for $i = 1, \ldots, N$.

The proof of this lemma is an inductive argument on n, which makes use of Theorem 1.9(iv), together with the properties of $\mathscr{L}$ and $\langle , \rangle$ developed in section 2.3.

Let σ_t denote the matrix $(\langle \xi_i(t), \xi_j(t) \rangle)_{i,j=1}^d$, for $t \geq 0$. Note that we can write σ_t in the form

$$\sigma_t = \int_0^t X(s, t) A(\xi_s) A(\xi_s)^* X(s, t)^* \, ds \tag{7.10}$$

where $X(s, t)$ satisfies the equation

$$X(s, t) = I + \sum_{k=1}^d \int_s^{t \vee s} [S_k(u) X(s, u) + X(s, u) S_k(u)^*] \, dw_k(u)$$
$$+ \int_s^{t \vee s} \left(\sum_{k=1}^d S_k(u) X(s, u) S_k^*(u) + C(u) X(s, u) \right.$$
$$\left. + X(s, u) C(u)^* \right) du$$

where the matrices $S_k(u)$, $k=1,\ldots,d$ and $C(u)$ are as defined in (2.25) (in fact equations (4.7) and (4.8) give σ_t in this form). It follows from (7.10) and (7.8) that

$$\sigma_t \geq \varepsilon \int_0^t X(s,t)\,\mathrm{d}s. \tag{7.11}$$

In view of this, a lower bound on the determinant of σ_t will follow from corresponding estimates for the matrices $\{X(s,t): s \leq t\}$. The following result will be used to obtain these.

Lemma 7.6 *Let $U_k:[0,\infty)\times\Omega\to\mathbf{M}^{d,d}$, $1\leq k\leq d$ and $V:[0,\infty)\times\Omega\to\mathbf{R}^d$ be adapted functions such that*

$$\begin{aligned} &\max_{1\leq i,j,k\leq d}\ \sup_{(t,\omega)} \|(U_k)_{ij}(t,\omega)\| \leq K \\ &\max_{1\leq i,j\leq d}\ \sup_{(t,\omega)} \|V_{ij}(t,\omega)\| \leq K \end{aligned} \tag{7.12}$$

for some finite constant K.

Suppose also that $(U_k)_{i,j}\equiv 0$ if either $|i-k|>L$ or $|j-k|>L$ and

$$V_{ij}\equiv 0 \quad \text{if} \quad |i-j|>L.$$

Suppose that M_0 is a deterministic non-negative symmetric matrix with trace ≤ 1 and for a given $s\geq 0$ define a matrix process $M(\cdot)$ by

$$\begin{aligned} M_t = M_0 &+ \sum_{k=1}^{d} \int_s^{t\vee s} (U_k M + M U_k^*)_u\,\mathrm{d}w_k(u) \\ &+ \int_s^{t\vee s} \left(\sum_{k=1}^{d} U_k M U_k^* + VM + MV^* \right)_u \mathrm{d}u. \end{aligned} \tag{7.13}$$

Then for each t, M_t is symmetric and non-negative definite. Furthermore, given any $1\leq p<\infty$ and $T>0$, there exists a constant θ, depending only on p, T, K, L such that for all $t\leq T$

$$E[|\mathrm{Tr}\, M_t|^p] \leq \theta \tag{7.14}$$

where Tr *denotes the trace of a matrix.*

Proof Note that we may assume, without loss of generality, that $s=0$. The symmetry and non-negative definiteness of M_t follows from the fact that $M_t = N_t N_t^*$, where

$$N_t = M_0^{1/2} + \sum_{k=1}^{d} \int_0^t U_k N_u\,\mathrm{d}w_k(u) + \int_0^t V N_u\,\mathrm{d}u.$$

Let $T_t = \operatorname{Tr} M_t$. Define

$$\gamma_k(t) = \operatorname{Tr}(U_k M_t)/T_t$$
$$\beta(t) = \operatorname{Tr}\left(\left(VM + MV^* + \sum_{k=1}^{d} U_k M U_k^*\right)_t\right)\Big/T_t.$$

Since Tr is a linear operator we will have

$$T_u = T_0 + 2\sum_{k=1}^{d}\int_0^t \gamma_k(u) T_u \, dw_k(u) + \int_0^t \beta(u) T_u \, du$$

where $T_0 = \operatorname{Tr} M_0$.

Solving this equation for T_t gives

$$T_t = T_0 \exp\left\{2\sum_{k=1}^{d}\int_0^t \gamma_k(u)\, dw_k(u) + \int_0^t \left[\beta(u) - 2\sum_{k=1}^{d}\gamma_k^2(u)\right] du\right\}. \tag{7.15}$$

Using condition (7.12) one can obtain, via elementary manipulations, the inequalities

$$|\beta(t)|, \qquad \sum_{k=1}^{d}\gamma_k(t)^2 \le C_1 \tag{7.16}$$

where C_1 is a constant depending only upon K and L.

From (7.15) we obtain

$$E[T_t^p)] = E[Q_t J_t]$$

where

$$Q_t = T_0^p \exp\left\{2p\sum_{k=1}^{d}\int_0^t \gamma_k(u)\, dw_k(u) - 4p^2\int_0^t \sum_{k=1}^{d}\gamma_k(u)^2\, du\right\}$$

$$J_t = \exp\left\{(4p^2 - 2p)\int_0^t \sum_{k=1}^{d}\gamma_k(u)^2\, du + p\int_0^t \beta(u)\, du\right\}$$

Inequality (7.16) implies that for $t \le T$, J_t is bounded by a constant depending only on T, p, K and L. Furthermore the Girsanov theorem implies that $E[Q_t] = T_0^p$. Thus (7.14) follows. ■

The next result is a consequence of Lemma 7.6.

Lemma 7.7 *Let U_k, $1 \le k \le d$, V and M be as in Lemma 7.6 with $M_0 = I$. Then given $1 \le N \le d$, $1 \le p \le \infty$ and $T > 0$, there exists a constant*

θ, depending only on p, N, T, K and L such that for all $t \le T$

$$E\left[\left|\sum_{i=1}^{N} M_{ii}(t)\right|^p\right] \le \theta.$$

In order to derive this, apply Lemma 7.6 with $\hat{M} \equiv \Lambda M \Lambda$ in place of M, $\hat{U}_k \equiv \Lambda U_k \Lambda^{-1}$ in place of U_k, and $\hat{V} \equiv \Lambda U_k \Lambda^{-1}$ in place of V where Λ is the matrix

$$\left[\frac{1}{2^j}\delta_{ij}\right]_{i,j=1}^{d}.$$

The lemma follows from the fact that

$$\sum_{i=1}^{N} M_{ii} \le 4^N \sum_{k=1}^{d} \hat{M}_{ii}.$$

Lemma 7.8 *Let $1 \le k \le d$ and suppose that*

$$A = \begin{bmatrix} A_{(11)} & A_{(12)} \\ A_{(21)} & A_{(22)} \end{bmatrix}$$

is a symmetric positive definite matrix of dimension $d \times d$, where $A_{(11)}$, $A_{(12)}$, $A_{(21)}$ and $A_{(22)}$ are of dimensions $k \times k$, $(d-k) \times k$, $k \times (d-k)$ and $(d-k) \times (d-k)$ respectively. Then

$$(A_{(11)})^{-1} \le (A^{-1})_{(11)}$$

where $(A^{-1})_{(11)}$ denotes the principal $k \times k$ minor of A^{-1}.

The proof of this lemma is elementary. We are now in a position to prove Theorem 7.4.

Let $X(s, t)$ be as in (7.10). Let $Z(s, t) = X(s, t)^{-1}$. By Itô's lemma we have

$$\begin{aligned} Z(s, t) = I - \sum_{k=1}^{d} \int_s^{t \vee s} [S_k^*(u)Z(s, u) + Z(s, u)S_k(u)]\, dw_k(u) \\ + \int_s^{t \vee s} \Big\{ -(C^*(u)Z(s, u) + Z(s, u)C(u)) \\ + \sum_{k=1}^{d} (S_k^{*2}(u)Z(s, u) + Z(s, u)S_k^2(u)) \Big\}\, ds. \end{aligned}$$

Let $X^{(N)}(s, t)$ denote the principal $N \times N$ minor of $X(s, t)$. It follows from Lemma 7.8 that

$$\mathrm{Tr}(X^{(N)}(s, t))^{-1} \le \sum_{i=1}^{N} Z_{ii}(s, t).$$

Lemma 7.7 now implies that for each $T>0$ and $1\leq p<\infty$, there exists a constant $C_p(T, N)$ depending only on N, T, p, L, K and the bounds on the first derivatives of A and B such that

$$E[|\mathrm{Tr}\, X^{(N)}(s, t)^{-1}|^p]\leq C_p(T, N) \tag{7.17}$$

for all $0\leq s,\ t\leq T$.

Let $\sigma_t^{(N)}\equiv(\langle\xi_i(t), \xi_j(t)\rangle)_{i,j=1}^N$ as defined earlier. In view of (7.11) we have

$$\sigma_t^{(N)}\geq\varepsilon\int_0^t X^{(N)}(s, t)\, ds\geq\varepsilon\int_0^t (\mathrm{Tr}(X^{(N)}(s, t)^{-1}))^{-1}\, dsI \tag{7.18}$$

Let $\Delta_t^{(N)}$ denote det $\sigma_t^{(N)}$. Applying Jensen's inequality to (7.18) gives

$$\begin{aligned}|\Delta^{(N)}(t)|^{-p} &\leq\frac{1}{\varepsilon^{Np}}\left(\int_0^t (\mathrm{Tr}(X^{(N)}(s, t)^{-1}))^{-1}\, ds\right)^{-Np}\\ &\leq\frac{t^{-Np-1}}{\varepsilon^{Np}}\int_0^t ((\mathrm{Tr}\, X^{(N)}(s, t))^{-1})^{Np}\, ds.\end{aligned}$$

Using (7.17) we obtain

$$E[|\Delta^{(N)}(t)|^{-p}]\leq\left(\frac{1}{\varepsilon t}\right)^{Np} C_{Np}(T, N)\qquad\text{for all } 0<t\leq T. \tag{7.19}$$

Let ν_N denote the distribution of $(\xi_1, \ldots, \xi_N)$ on $\mathbf{R}^N$. Recall that the methods of chapter 2 yield the following result: for every natural number b and vectors $y_1, \ldots, y_b$ in $\mathbf{R}^N$, there exists an L^1 random variable X such that for all test functions ϕ on $\mathbf{R}^N$

$$\int_{\mathbf{R}^d} D^b\phi(x)(y_1, \ldots, y_b)\, d\nu_N(x)=\int_{C_0(\mathbf{R}^d)}\phi(w)X(w)\, d\gamma(w)$$

where X is an algebraic expression in $\xi^{(b)}(\xi_1, \ldots, \xi_N)$ and $(\Delta_t^{(N)})^{-1}$. It follows from Lemma 7.5 and (7.19) that

$$\left|\int_{\mathbf{R}^d} D^b\phi(x)(y_1, \ldots, y_b)\, d\nu_N(x)\right|\leq C\,\|\phi\|_\infty$$

where C depends on $b, y_1, \ldots, y_b, t, N, k$ and L but is independent of $d\geq N$. Theorem 7.4 now follows from Lemma 1.14.

7.3 Towards the construction of quasi-invariant measures on an infinite dimensional vector space

Let E be a Banach space such that E and E^* are both separable. A Borel measure ν defined on E is said to be *quasi-invariant* under a subspace K

if the class of its null sets is invariant under translation by every element of K. A Gaussian measure is quasi-invariant under its canonical Hilbert subspace, by the Cameron–Martin theorem. If ν is an arbitrary Borel measure on a Euclidean space $\mathbf{R}^n$, then quasi-invariance of ν with respect to the whole of $\mathbf{R}^n$ implies equivalence to the Lebesgue measure (cf. Bell[3]). This result has no analogue in infinite dimensions; nonetheless the quasi-invariance properties of a measure on an infinite dimensional vector space contain important information about the structure of the measure.

The results that appear in this section represent ongoing research of the author. They are part of a programme for studying quasi-invariance of the measures induced by certain stochastic differential equations with values in an infinite dimensional Hilbert space. In order to outline our programme we need to introduce the following criterion of measure differentiability. A definition of this type was first given by Kuo[16].

Definition 7.9 (after Bell[2]) ν is *differentiable* with respect to a subspace K of E if, for every vector r in K, there exists an L^1 random variable X_r such that the following relation holds for every test function† φ

$$\int_E D\varphi(x)r\,\mathrm{d}\nu(x) = \int_E \varphi(x)X_r(x)\,\mathrm{d}\nu(x). \tag{7.20}$$

That a relationship exists between this property and quasi-invariance is illustrated by the next result which is proved in Bell[2]. See also the work of Kats[15].

Theorem 7.10 (after Bell[2]) Suppose that ν is differentiable with respect to the line L spanned by a vector $r \in E$ and the function X_r in (7.20) *satisfies the following conditions*:

(i) *$X_r(x + \cdot)$ is continuous as a function on L for almost all x.* (7.21)

(ii) $\sup_{t\in[0,1]} [X_r(x + tr)]^4$

and

$$\sup_{s,t\in[0,1]} \exp\left\{-4\int_s^t X_r(x + ur)\,\mathrm{d}u\right\}$$

are locally integrable. Then ν is quasi-invariant under translation by r and,

† Here a test function means a bounded C^1 function from E into $\mathbf{R}$ with a bounded first derivative.

denoting by ν_r *the measure* $\nu(\cdot + r)$, *we have*

$$\frac{d\nu_r}{d\nu}(x) = \exp\left\{ -\int_0^1 X_r(x+ur)\, du \right\} \quad a.s.$$

In this section the methods of chapter 4 will be used to show that under certain conditions a stochastic differential equation with values in a Hilbert space E induces measures on E which are differentiable with respect to a dense linear subspace K. We would like to be able to infer from Theorem 7.10 that these measures are quasi-invariant under K; unfortunately at the time of writing we are unable to verify that the respective random variables $\{X_r : r \in K\}$ in (7.20) satisfy condition (7.21). However, we are optimistic that it will be possible to use Theorem 7.13 of this section, in conjunction with a modified form of Theorem 7.10 to derive the required quasi-invariance result and that this will yield a rich supply of (non-Gaussian) quasi-invariant measures on E.

Suppose that (i, H, E) is an abstract Wiener space with Gaussian measure μ on E. Assume that E is a Hilbert space (note that this implies i is Hilbert–Schmidt). As usual we embed E^* inside E via the maps

$$E^* \xrightarrow{i^*} H^* \cong H \xrightarrow{i} E.$$

Denote the embedded subspace E^* by K, and give K the (Hilbert) norm induced from E^*. Let $w : \Omega \times [0, 1] \to E$ be a corresponding Brownian motion. This is a continuous, time-homogeneous, independent increment stochastic process in E with initial point 0 such that $w(1)$ has distribution μ. The law of the process w, which we will denote by γ, is the Gaussian measure arising from the abstract Wiener space $(j, L_0^{2,1}(H), C_0(E))$, where $C_0(E)$ denotes the Banach space of paths σ from $[0, 1]$ into E with $\sigma(0) = 0$ under the norm $\|\sigma\| = \sup\{\|\sigma_t\|_E : t \in [0, 1]\}$; $L_0^{2,1}(H)$ the subspace of E consisting of absolutely continuous paths into H with square integrable derivatives, equipped with the inner product

$$\langle \sigma, \tau \rangle = \int_0^1 \langle \sigma_t', \tau_t' \rangle_H \, dt; \qquad \sigma, \tau \in L_0^{2,1}(H)$$

and j the inclusion map of $L_0^{2,1}(H)$ into $C_0(E)$.

For any Hilbert spaces P and Q, denote by $L_2(P, Q)$ the space of Hilbert–Schmidt operators from P to Q with the Hilbert–Schmidt norm. Let $A : E \to L(E)$ be a bounded C^2 map such that DA and D^2A are both bounded. Finally let $x \in E$ and $t \in (0, 1]$ be fixed. With these preliminaries we are now ready to introduce the infinite dimensional stochastic differential equation to which we alluded above.

It can be shown that the following equation has a unique adapted continuous solution with values in E (cf. Metivier and Pellaumail[23]).

$$\xi_s = x + \int_0^t A(\xi_u)\,dw_u, \qquad s \in [0, 1]. \tag{7.22}$$

Let ν denote the law of ξ_t.

The main result of this section is Theorem 7.13. The next lemma is an essential ingredient in its proof.

Lemma 7.11 *Suppose that A satisfies the following additional conditions:*
(i) $A: E \to L(H, K)$, $DA: E \to L^2(H, E; K)$, $D^2A: E \to L^3(H, H, E; K)$ *boundedly and continuously (here L^r denotes r–linear map).*
(ii) *There exists $\varepsilon > 0$ such that $\langle A(y)A(y)^*h, h\rangle_K \geq \varepsilon \|h\|_K^2$ $\forall y \in E$, $h \in K$, where $A(y)^*$ denotes the adjoint of $A(y)$ as an operator from H into K.*

Then for every $s \in [0, 1)$ the following operator–valued equation has a unique continuous adapted solution with values in the affine space $I + L_2(K, K)$:

$$Z_v^{(s)} = I + \int_s^v DA(\xi_u)(Z_u^{(s)}, dw_u), \qquad v \in [s, 1]. \tag{7.23}$$

Furthermore if σ denotes the matrix

$$\int_0^t Z_t^{(s)} A(\xi_s) A(\xi_s)^* Z_t^{(s)*}\,ds$$

then $\sigma \in GL(K)$ and $\|\sigma^{-1}\|_{L(K)} \in L^p$ for all p.

The proof of Lemma 7.11 will require

Lemma 7.12 *If $T \in L(K)$ is a non-negative definite self-adjoint operator and $m = \inf_{\|h\|=1} \langle Th, h\rangle \neq 0$, then $T \in GL(K)$ and $\|T^{-1}\| \leq 1/m$.*

Proof Define $M = \sup_{\|h\|=1} \langle Th, h\rangle$. By the spectral theorem we have

$$T = \int_m^{M+\varepsilon} \lambda\,dE_\lambda \qquad \text{for any } \varepsilon > 0.$$

Define an operator $B \in L(K)$ by

$$B = \int_m^{M+\varepsilon} 1/\lambda\,dE_\lambda$$

Then $TB = \int_m^{M+\varepsilon} 1\,dE_\lambda = E_{M+\varepsilon} - E_m = I$, thus $T \in GL(K)$ and $B = T^{-1}$.

Furthermore, we have $\|B\| \leq \sup_{m \leq \lambda \leq M} 1/\lambda = 1/m$ (see, for example, "Foundations of Modern Analysis" by A. Friedman, Theorem 6.75). ■

Proof of Lemma 7.11 Condition (i) implies that $y \to DA(y)$ is a bounded, continuous map from E into $L_2(K \otimes H, K)$. The first part of the lemma follows from this, and the fact that the set of Hilbert–Schmidt operators on K form an ideal in $L(K)$.

Lemma 7.12 and condition (ii) give

$$
\begin{aligned}
1/\|\sigma^{-1}\| &\geq \inf_{\|h\|=1} \langle \sigma h, h \rangle_K \\
&= \inf_{\|h\|=1} \int_0^t \|A(\xi_s)^* Z_t^{(s)*} h\|_H^2 \, ds \\
&\geq \inf_{\|h\|=1} \epsilon \int_0^t \|Z_t^{(s)*} h\|_K^2 \, ds
\end{aligned} \tag{7.24}
$$

As usual we may use Itô's lemma to write down an integral equation for the process $Z^{(s)-1}$ and then standard estimates give $\sup \{E \|Z_t^{(s)-1}\|_{L(K)}^p : 0 \leq s \leq t \leq 1\} < \infty$ for all p. Applying this to (7.24) together with Jensen's inequality, we obtain

$$
\begin{aligned}
E \|\sigma^{-1}\|^p &\leq 1/\varepsilon^p \left[\inf_{\|h\|=1} \int_0^t \|Z_t^{(s)*} h\|_K^2 \, ds \right]^{-p} \\
&\leq 1/\varepsilon^p E \left[\int_0^t \inf_{\|h\|=1} \|Z_t^{(s)*} h\|_K^2 \, ds \right]^{-p} \\
&\leq 1/(\varepsilon t)^p E \left[1/t \int_0^t \left[\inf_{\|h\|=1} \|Z_t^{(s)*} h\|_K^2 \right]^{-p} ds \right. \\
&= 1/\varepsilon^p t^{p+1} \int_0^t E \|Z_t^{(s)^{-1}}\|_{L(K)}^{2p} \, ds < \infty
\end{aligned}
$$

Thus the lemma is proved. ■

Theorem 7.13 *Suppose the hypotheses of Lemma* 7.11 *are satisfied. Then v is differentiable with respect to K.*

The proof of this result follows similar lines to that of Theorem 4.9. Defining $\{g^m : m = 1, 2, \ldots\}$ on $C_0(E)$ as in (4.2) and $\sigma_m = Dg_t^m(w) Dg_t^m(w)^*$, we will have $\sigma_m \to \sigma$ in $L(K)$ a.s., where σ is as in

Lemma 7.11. It follows that $\sigma_m^{-1} \to \sigma^{-1}$ in $L(K)$ a.s. Let $r \in K$ and ϕ be any test function. Arguing as in chapter 4 we obtain the equality

$$\int_E D\phi(y)r\,dv(y) = \int_{C_0(E)} \phi(g_t(w))Y_r(w)\,d\gamma(w)$$

$$= \int_E \phi(y)E[Y_r(w)/\xi_t = y]\,dv(y)$$

where $Y_r(w)$, the limit in $L^2(\gamma)$ of the sequence $\{\mathrm{Div}[Dg_t^m(w)^*(\sigma_m^{-1}r)]\}$, is given by

$$Y_r(w) = \langle \sigma(w)^{-1}r, \eta_t \rangle$$

$$-\sum_n \{\langle \sigma(w)^{-1}r, D^2g_t(w)(f_n, f_n)\rangle$$

$$-\langle \sigma(w)^{-1}D\sigma(w)f_n\sigma(w)^{-1}r, Dg_t(w)f_n\rangle\}.$$

Here $\langle , \rangle$ denotes the natural pairing of E and E^* and η the solution of the equation

$$\eta_s = \int_0^s A(\xi_u)\,dw_u + \int_0^s DA(\xi_u)(\eta_u, dw_u), \qquad s \in [0, 1].$$

$\{f_n\}$ denotes any orthonormal basis of $L_0^{2,1}(H)$ and $Dg_t(w)$, $D^2g_t(w)$ and $D\sigma(w)$ are the maps obtained by formally differentiating with respect to w in equations (7.22) and (7.23).

Thus the theorem holds.

Concluding Remark The above hypotheses can be satisfied if E is infinite dimensional. However, the condition that A maps into $L(E)$, required for the definition of equation (7.22), implies that for each $y \in E$ $A(y)A(y)^*$ is a Hilbert–Schmidt operator on E†. In particular condition (ii) in Lemma 7.11 will not be satisfied with respect to the E-inner product and E-norm in the infinite dimensional case, and it is therefore not possible to extend the foregoing argument to prove differentiability of v with respect to E. This limitation reflects the fact that a non-zero Borel measure on an infinite dimensional Banach space cannot be differentiable in all directions.

† Where * denotes adjoint between the spaces H and E.

References

1. Bell D 1982 Some properties of measures induced by solutions of stochastic differential equations. PhD thesis, Univ Warwick.
2. Bell D 1985 A quasi-invariance theorem for measures on Banach spaces, *Trans Amer Math Soc* **290** no. 2: 851–5.
3. Bell D 1986 On the relationship between differentiability and absolute continuity of measures on $\mathbf{R}^n$, *Prob Th Rel Fields* **72:** 417–24. Springer-Verlag, Berlin and New York.
4. Bichteler K 1981 Stochastic integrators with stationary independent increments, *Z Wahrsch Verw Gebiete* **58:** 529–48.
5. Bichteler K and Fonken D 1983 A simple version of the Malliavin calculus in dimension *N, Seminar on Stochastic Processes* (1982): 97–110. Birkhauser, Boston.
6. Bismut JM 1981 Martingales, the Malliavin calculus and hypoellipticity under general Hörmander's conditions, *Z Wahrsch Verw Gebiete* **56:** 469–505.
7. Elworthy KD 1974 Gaussian measures on Banach spaces and manifolds, *Global Analysis and Applications,* Vol. 2, pp 151–66. International Atomic Energy Agency, Vienna.
8. Elworthy KD 1982 *Stochastic Differential Equations on Manifolds.* Cambridge University Press, Cambridge and New York.
9. Friedman A 1975 *Stochastic Differential Equations and Applications,* Vol. 1. Academic Press, New York.
10. Gaveau B and Trauber P 1982 L'intégrale stochastique comme opérateur de divergence dans l'espace fonctionnel, *J Funct Anal* **46:** 230–8.
11. Gikhman II and Skorohod AV 1972 *Stochastic Differential Equations.* Springer-Verlag, Berlin and New York.
12. Gross L 1965 Abstract Wiener spaces, *Proc Fifth Berkeley Sympos Math Statist and Probability* Vol. 2 part 1: 31–42.
13. Holley R and Stroock D 1981 Diffusions on an infinite dimensional torus, *J Funct Anal* **42** no. 1: 29–63.
14. Hörmander L 1967 Hypoelliptic second order differential equations, *Acta Math* **119:** 147–71.
15. Kats MP 1978 Quasi-invariance and differentiability of measures, *Communications of the Moscow Math Soc*: 159.
16. Kuo H-H 1974 Differentiable measures, *Chinese Journal of Math* **2:** 189–99.
17. Kusuoka S 1982 Analytic functionals of Wiener processes and absolute

continuity, *Functional Analysis in Markov Processes* (Katato/Kyoto 1981) pp 1–46, Lecture Notes in Mathematics, no. 923. Springer-Verlag, Berlin and New York.

18. Kusuoka S and Stroock D 1985 Applications of the Malliavin calculus, Part II, *Journal of the Faculty of Science, Univ Tokyo* **32** no. 1: 1–76.
19. Malliavin P 1976 Stochastic calculus of variations and hypoelliptic operators, *Proceedings of the International Conference on Stochastic Differential Equations, Kyoto* pp 195–263. Kinokuniya, Tokyo; Wiley, New York.
20. Malliavin P 1978 C^k-hypoellipticity with degeneracy. In Friedman A and Pinsky M (eds). *Stochastic Analysis* pp. 199–214, 327–340. Academic Press, New York and London.
21. McShane EJ 1974 *Stochastic Calculus and Stochastic Models.* Academic Press, New York.
22. Michel D 1981 Régularité des lois conditionnelles en théorie du filtrage non-lineaire et calcul des variations stochastique, *J Funct Anal* **41** no. 1: 8–36.
23. Metivier M and Pellaumail J 1980 *Stochastic Integration* Academic Press, New York.
24. Norris J 1985 Simplified Malliavin calculus, *Seminaire de Probabilités* XIX.
25. Ramer R 1974 On nonlinear transformations of Gaussian measures, *J Funct Anal* **15:** 166–87.
26. Stroock D 1981 The Malliavin calculus and its applications, *Stochastic Integrals* (Proc. Sympos. Univ. Durham, Durham, 1980) pp 394–432, Lecture Notes in Mathematics, no. 851. Springer-Verlag, Berlin and New York.
27. Stroock D 1981 The Malliavin calculus and its application to second order parabolic differential equations I, *Math Systems Theory* **14** no. 2: 25–65.
28. Stroock D The Malliavin calculus and its application to second order parabolic differential equations II, *Math Systems Theory* **14** no. 2: 141–71.
29. Zakai M 1984 *The Malliavin calculus,* Preprint.

Index

Appendix: Admissible vector fields and quasi-invariant measures

This appendix, which contains recent work of the author, can be read independently of the preceding material. We use the scheme outlined in Section 5.3 to study the transformation of measure under the flow of a vector field.

Let X denote either a Banach space or a compact smooth finite-dimensional manifold without boundary, equipped with a finite Borel measure γ. Let Z be a C^1 vector field on X and consider the flow $x \mapsto \sigma_s$ generated by Z, defined by the differential equation

$$\dot{\sigma}_s = Z(\sigma_s), \quad \sigma_0 = x.$$

Assume the flow exists for all time s and defines a C^1 function in x. This is automatic in the compact manifold case and holds in the Banach space case provided Z satisfies the global Lipschitz condition.

$$||Z(x) - Z(y)|| \leq c||x - y||, \quad \forall x, y$$

for some constant c.

Definition A.1. *The vector field Z is said to be admissible if there exists a random variable Y such that the relation*

$$\int_X D\Phi(x) Z d\gamma = \int_X \Phi(x) Y d\gamma$$

holds for all test functions Φ on X. The random variable Y is called the divergence of Z and will be denoted by $Div(Z)$ in the sequel.

Example Let $Z \in K$, where K is the space in Section 7.3 and let ξ denote the solution process to the stochastic differential equation (7.22). Then the conclusion of Theorem 7.13 asserts that Z is admissible with respect to the law of ξ_t, for every $t \in (0, 1]$. We note that the argument in Section 7.3 and Theorem 7.13 easily extend to the case where Z is a C^1 map from E into K, with Z and DZ bounded.

Definition A.2. *We say that γ is quasi-invariant (under the flow σ_s) if the measures γ and $\gamma_s \equiv \sigma_s(\gamma)$ are equivalent, for all s.*

In Bell[2], the author introduced a method for studying the quasi-invariance of a measure γ defined on a Banach space E, under translation by a vector $h \in E$. The argument entailed studying the quasi-invariance of γ under the family of translations $\{sh, s \in \mathbf{R}\}$, i.e. the flow generated by the (constant) vector field h on E. In this appendix, both the method and the result in Bell[2] will be generalized to the case of non-constant vector fields. The main result is Theorem A.8. We begin with the following

Theorem A.3. *Suppose γ is quasi-invariant and the family of Radon-Nikodym derivatives $X_s \equiv d\gamma_s/d\gamma$ are differentiable in s. Suppose furthermore the random variables ZX_s are admissible. Then X_s satisfies the differential equation*

$$X_s' = Div(ZX_s). \qquad (A.$$

Proof Let Φ be a test function on X. Then

$$\int_X \Phi \circ \sigma d\gamma = \int_X \Phi X_s d\gamma.$$

Replacing Φ by $\Phi \circ \sigma_s^{-1}$ we have

$$\int_X \Phi \circ \sigma_s^{-1} X_s d\gamma = \int_X \Phi d\gamma \qquad (A.$$

thus the left hand side is constant in s. Differentiating with respect to s gives

$$0 = \int_X \Big\{ D\Phi(\sigma_s^{-1}(x)) \frac{d}{ds}\sigma_s^{-1}(x) X_s + \Phi \circ \sigma_s^{-1} X_s' \Big\} d\gamma. \qquad (A.3$$

Differentiating with respect to s in $\sigma_s^{-1}(\sigma_s(x)) = x$, we have

$$\frac{d}{ds}\sigma_s^{-1}(\sigma_s(x)) + D\sigma_s^{-1}(\sigma_s(x))\dot{\sigma}_s(x) = 0$$

thus

$$\frac{d}{ds}\sigma_s^{-1}(\sigma_s(x)) + D\sigma_s^{-1}(\sigma_s(x)) Z(\sigma_s(x)) = 0$$

and we obtain

$$\frac{d}{ds}\sigma_s^{-1}(x) = -D\sigma_s^{-1}(x) Z(x).$$

Substituting this into (A.3) gives

$$0 = \int_X \Big\{ - D\Phi(\sigma_s^{-1}(x)) D\sigma_s^{-1}(x) Z(x) X_s + \Phi \circ \sigma_s^{-1} X_s' \Big\} d\gamma$$
$$= \int_X \Big\{ - D(\Phi \circ \sigma_s^{-1})(x)(ZX_s) + \Phi \circ \sigma_s^{-1} X_s' \Big\} d\gamma.$$

Thus

$$\int_X \Phi \circ \sigma_s^{-1} \Big\{ X_s' - Div(ZX_s) \Big\} d\gamma = 0. \qquad (A.4$$

Since this holds for all test functions Φ, we conclude that (A.1) holds.

Theorem A.4. *Suppose Z is admissible and there exist a family of random variables X_s with $X_0 = 1$ satisfying (A.1). Then γ is quasi-invariant under the flow generated by Z and*

$$\frac{d\gamma_s}{d\gamma}(x) = X_s(x) = \exp\Big\{\int_0^s Div(Z)(\sigma_{-u}(x))du\Big\}.$$

The proof will require the following

Lemma A.5. *Define the pull back $\sigma_s^*(Z)$ of the vector field Z under σ_s by*

$$\sigma_s^*(Z)(x) = D\sigma_s(x)^{-1}Z(\sigma_s(x)). \tag{A.5}$$

Then $\sigma_s^*(Z) = Z$, i.e. Z is invariant under σ_s.

Proof We have

$$\begin{aligned} 0 &= \frac{d}{ds}\sigma_s^{-1}(\sigma_s(x)) \\ &= (\frac{d}{ds}\sigma_s^{-1})\sigma_s(x) + D\sigma_s^{-1}(\sigma_s(x))\dot{\sigma}_s(x) \\ &= (\frac{d}{ds}\sigma_s^{-1})\sigma_s(x) + D\sigma_s^{-1}(\sigma_s(x))Z(\sigma_s(x)). \end{aligned}$$

Substituting this into (A.5) gives

$$\begin{aligned} \sigma_s^*(Z)(x) &= -(\frac{d}{ds}\sigma_s^{-1})\sigma_s(x) \\ &= (\frac{d}{ds}\sigma_{-s})(\sigma_s(x) \\ &= Z(x) \end{aligned}$$

as required.

We are now in a position to prove Theorem A.4. Assume there exist a family of random variables X_s satisfying $X_0 = 1$ and $X_s' = Div(ZX_s)$, i.e. (A.4) holds. Reversing the argument used to prove Theorem A.3, we deduce from (A.4) that (A.2) holds for all test functions Φ. This implies that γ_s is equivalent to γ and

$$\frac{d\gamma_s}{d\gamma} = X_s. \tag{A.6}$$

Let Φ be an arbitrary test function on X. Using (A.6) and Lemma A.5, we have

$$\int_X \Phi Div(ZX_s)d\gamma = \int_X D\Phi(x)(ZX_s)d\gamma$$

$$= \int_X D\Phi(x)Z.X_s d\gamma$$

$$= \int_X D\Phi(\sigma_s(x))Z(\sigma_s(x))d\gamma$$

$$= \int_X D(\Phi \circ \sigma_s)(x)\sigma_s^*(Z)d\gamma$$

$$= \int_X D(\Phi \circ \sigma_s)(x)Zd\gamma$$

$$= \int_X \Phi \circ \sigma_s(x)Div(Z)d\gamma = \int_X \Phi Div(Z)(\sigma_s^{-1}(x))X_s d\gamma.$$

Thus we obtain the key relation

$$Div(ZX_s)(x) = X_s Div(Z)(\sigma_s^{-1}(x)).$$

Substituting this into (A.1) gives

$$X_s' = X_s Div(Z)(\sigma_{-s}).$$

It follows that

$$X_s = \exp\left\{\int_0^s Div(Z)(\sigma_{-u})du\right\}$$

which yields the result.

Definition A.6. *We say that a function $F : X \mapsto \mathbf{R}$ is Z-differentiable at $x \in X$ if $\rho(s) \mapsto F(\sigma_s(x))$ is differentiable at $s = 0$ and denote*

$$\rho'(0) = DF(x)Z(x).$$

We give another method for deriving the formula for the Radon-Nikoym derivatives in Theorem A.4.

Lemma A.7. *Suppose Z is admissible and $F : X \mapsto \mathbf{R}$ is Z- differentiable. Then the vector field ZF is admissible and*

$$Div(ZF) = FDiv(Z) - DF(x)Z.$$

Proof For any test function Φ on X, we have

$$\begin{aligned}\int_X \Phi F Div(Z) d\gamma &= \int_X D(\Phi F)(x) Z d\gamma \\ &= \int_X \{F d\Phi(x) Z + \Phi DF(x) Z\} d\gamma \\ &= \int_X \{d\Phi(x)(FZ) + \Phi DF(x) Z\} d\gamma \\ &= \int_X \Phi\{Div(FZ) + DF(x) Z\} d\gamma.\end{aligned}$$

The lemma follows.

Suppose now the hypotheses of Theorem A.4 hold and the random variables $X_s(x)$ are Z-differentiable. Applying Lemma A.7 to equation (A.1) gives

$$X_s' = X_s Div(Z) - DX_s(x)Z. \tag{A.7}$$

Define

$$Y_s(x) = X_s(\sigma_s(x)).$$

Then

$$\begin{aligned}Y_s' &= X_s'(\sigma_s(x)) + DX_s(\sigma_s(x))\dot{\sigma}_s(x) \\ &= X_s'(\sigma_s(x)) + DX_s(\sigma_s(x))Z(\sigma_s(x))\end{aligned}$$

which, using (A.7)

$$\begin{aligned}&= X_s(\sigma_s(x)) Div(Z)(\sigma_s(x)) \\ &= Y_s Div(Z)(\sigma_s(x)).\end{aligned}$$

Together with the condition $Y_0 \equiv 1$, this yields

$$Y_s = \exp\left\{\int_0^s Div(Z)(\sigma_u) du\right\}.$$

Thus

$$\begin{aligned}X_s &= \exp\left\{\int_0^s Div(Z)(\sigma_u \circ \sigma^{-1}) du\right\} \\ &= \exp\left\{\int_0^s Div(Z)(\sigma_{u-s}) du\right\} \\ &= \exp\left\{\int_0^s Div(Z)(\sigma_{-u}) du\right\}.\end{aligned}$$

as before.

Theorem A.8. *Suppose Z is admissible and there exists $B \subseteq X$ with $\gamma_s(B) = 1$ for all s, such that $Div(Z)$ is defined and Z-differentiable on B. Suppose further that the function $s \mapsto Div(Z)(\sigma_s(x))$ is absolutely continuous for almost all $x \in B$. Then γ is quasi-invariant under Z and*

$$\frac{d\gamma_s}{d\gamma} = \exp\left\{\int_0^s (DivZ)(\sigma_{-u})du\right\}.$$

Proof We have

$$\frac{d}{du}(DivZ)(\sigma_{-u}(x)) = D(DivZ)(\sigma_{-u}(x)))\frac{d}{du}\sigma_{-u}(x))$$

$$= -D(DivZ)(\sigma_{-u}(x)))Z(\sigma_{-u}(x)))$$

which, by the invariance of Z (Lemma A.5)

$$= -D((DivZ) \circ \sigma_{-u})(x)Z(x).$$

Thus $(DivZ)\circ_{-u}$ is Z-differentiable and

$$D((DivZ) \circ \sigma_{-u})(x)Z(x) = -\frac{d}{du}(Div(Z))(\sigma_{-u}(x))).$$

Integrating with respect to s gives

$$(DivZ)(x) - Div(Z)(\sigma_{-s}(x)) = D\left[\int_0^s (DivZ)(\sigma_{-u})du\right](Z(x)). \quad \text{(A.8}$$

Define

$$X_s = \exp\left\{\int_0^s (DivZ)(\sigma_{-u})du\right\}.$$

By (A.8) and Lemma A.7, ZX_s is admissible and

$$Div(ZX_s)(x) = X_s DivZ - DX_s(x)Z$$

$$= X_s DivZ - X_s\left[DivZ - (DivZ)(\sigma_{-s}(x))\right]$$

$$= X_s(DivZ)(\sigma_{-s}(x))$$

$$= X_s'.$$

The result now follows from Theorem A.4.

Example Suppose (i, H, E) is an abstract Wiener space with Gaussian measure γ on E. Let $< .,. >_H$ and $|.|$ denote, respectively, the inner product and norm on H, and define $Z \equiv h$, where h is an element of H. Then

$$\sigma_s(x) = x + sh.$$

It can be shown (cf. eg. [L. Gross, Potential theory on Hilbert space, *J. Funct. Anal.* 1, 1967, pp. 123-181]) that Z is admissible and

$$DivZ(x) =< h, x >$$

where $< h, . >$ denotes a stochastic extension of the linear form$< h., . >_H$ to an L^2 random variable on E, defined by

$$< h, x >= \lim_n < h., P_n x >_H$$

where $\{P_n : E \mapsto H\}$ is any sequence of orthogonal projections converging strongly to the identity on E. Define $B \subseteq E$ to be the set on which the convergence holds. Then it is clear that B is *invariant* under σ_s and for $x \in B$,

$$< h, \sigma_s(x) >=< h, x > +s||h||^2.$$

In particular, $DivZ$ is Z-differentiable. Thus Theorem A.8 yields the

Cameron-Martin Theorem. *The abstract Wiener measure γ is quasi-invariant under translation by $h \in H$ and, denoting by γ_h the measure $\gamma(. - h)$, we have*

$$\frac{d\gamma_h}{d\gamma} = \exp\{ < h, x > -\frac{1}{2}||h||^2\}.$$